Fix Your Gut
Fix Your Life

Metagenomics and Microbiome

What Trillions of Gut Bacteria Are Doing to Your Health, Mood, and Future

C. Louis-Charles, PhD
Cybersoft Publishing LLC

Cybersoft Publishing LLC

Fort Washington, MD 20744

First Edition 2026

Medical disclaimer: This book is for general education only. It does not provide medical advice, diagnosis, or treatment. Consult a licensed clinician for personal health decisions.

Contents

Introduction

The envelope arrives on a Tuesday, mixed in with the electric bill and a pizza coupon. The stool test results, the ones you mailed off three weeks ago after your gastroenterologist mentioned that a commercial kit might help you "get a sense of things," are printed on four glossy pages.

The first page is colorful, almost cheerful: bar charts, pie charts, and a long ranked list of names you cannot pronounce. There is a column of numbers, each paired with a color-coded indicator. Some are green. Some are yellow. One, near the top, is red. Below the charts, a section titled "Your Microbiome Diversity Score" tells you that yours is "slightly below optimal." You are not sure what optimal means, who decided on it, or whether "slightly below" is the kind of thing you should call your doctor about or file away under "vaguely concerning."

You pour a cup of coffee. You read the report again. You still do not know what to do with it. That moment of confusion, equal parts fascination and helplessness, is exactly where this book begins.

What This Book Is About

Inside your gut right now, there are roughly as many microbial cells as human cells in your entire body. These bacteria, viruses, fungi, and other organisms are not passengers: they digest fibers your own enzymes cannot touch, manufacture vitamins, train your immune system, and send chemical signals that reach your brain. Collectively, they are called your microbiome (the full community of microorganisms living in and on you, along with their genes and metabolic products). For most of human history, scientists barely knew this community existed in its true complexity. Now, thanks to metagenomics (the practice of reading DNA directly from a biological sample without culturing individual microbes in a lab), we can finally see it.

What we are seeing is extraordinary. And also, in some corners, dramatically overhyped. That tension is what this book is about.

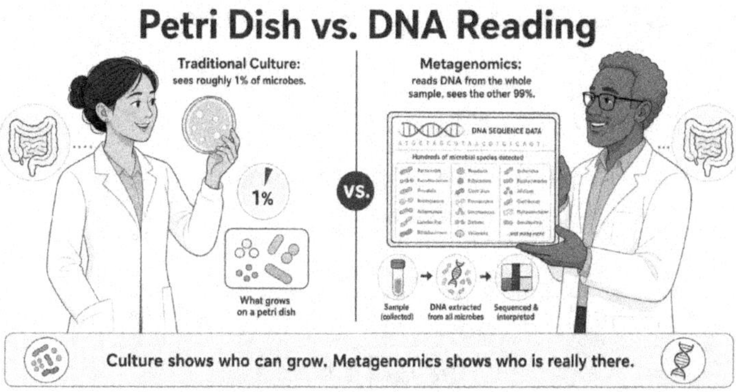

Diagram Intro.1. Petri Dish vs. DNA Reading

For most of the twentieth century, microbiology depended on culturing: you spread a sample on a nutrient plate, waited for colonies to grow, and studied whatever grew. The method gave us germ theory, vaccines, and antibiotics, but it had a hidden flaw. Most gut microbes, somewhere around ninety-nine percent, refuse to grow on a lab plate. The Petri dish was a filter that showed scientists only the loudest 1% of a vast conversation.

Metagenomics changed that. Instead of growing anything, metagenomics sequences all the DNA present in a sample, and every organism leaves behind its genetic signature. Sequencing (the laboratory process of reading the precise order of chemical letters, called nucleotide bases, in a strand of DNA) allows scientists to identify organisms entirely by their genetic code. It is something like identifying every musician in a symphony by analyzing the sound waves in a recording, without needing to see the stage.

This book will explain how that technology works, what it has revealed, and what it has yet to resolve.

Giving You the Tools to Improve Your Gut Life

The Book's Promise

This is not a wellness manifesto. It will not tell you that one probiotic capsule will transform your health, or that a stool test will reveal the hidden cause of your anxiety. Some of those claims are even partially supported by early evidence, but the words "partially" and "early" are doing enormous work, and this book will always say so plainly.

Every major claim here is grounded in peer-reviewed research. When the evidence is weak, inconsistent, or exists only in mouse studies not yet replicated in humans, you will hear that too. The book uses a three-tier evidence system throughout: Strongly Supported for claims backed by multiple consistent human trials, Promising for solid early evidence that needs more replication, and Marketing-Only for claims that have run well ahead of the data.

Diagram Intro.2. The Book's Journey Through Ten Stations

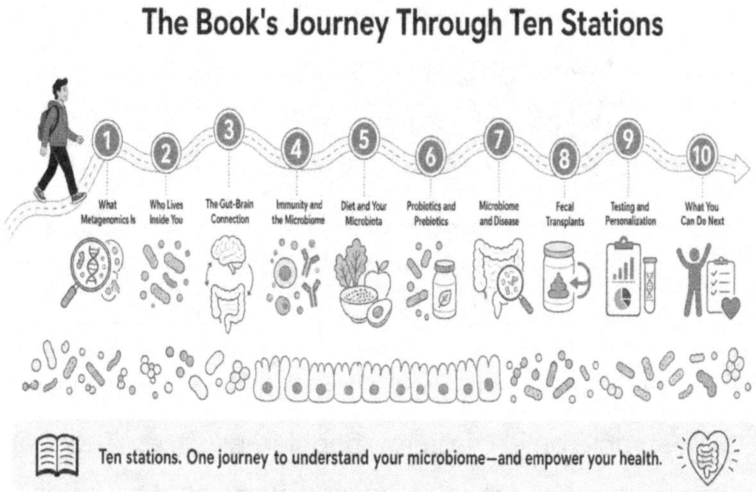

The path is not linear in the sense of being rigid. Each station stands on its own, and the journey is designed to feel cumulative rather than obligatory.

Early chapters are more methodological, explaining how scientists know what they claim to know. Later chapters are more applied. The evidence-tier system is the bridge between those two modes, and the diagram below shows exactly how it works.

Diagram Intro.3. How Confident Should You Be?

Those distinctions matter more in the microbiome field than in almost any other area of popular science. The gap between what headlines claim and what studies actually show is significant, and a central job of this book is to help you measure that gap so you can make good decisions with real information.

The Three Terms You Need From the Start

Three concepts will appear on nearly every page. You do not need a biology degree to understand them, but clear definitions help.

The microbiome is the full community of microorganisms living in and on your body, together with all their genes and metabolic products. The term is often used loosely to refer only to bacteria, but the technically complete version includes viruses, fungi, and other organisms. When this book uses "microbiome," it means the whole community unless otherwise specified.

Metagenomics is the approach scientists use to study a community without culturing individual organisms in a lab. Whereas classical genomics sequences the DNA of a single known organism, metagenomics sequences all the DNA in a sample at once and uses computational tools to sort the resulting fragments by species. In practical terms, it is the technology that made modern microbiome science possible.

Sequencing is the laboratory method used to determine the precise chemical sequence of DNA. Think of DNA as a long sentence written in a four-letter alphabet: adenine, thymine, cytosine, and guanine. Modern machines can read billions of those letters in a single run, which is why the cost has fallen from three billion dollars in 2003 to a few hundred dollars today. That price drop is also what made commercial stool testing possible.

The Chapters Ahead

The chapters ahead each cover a distinct piece of the microbiome story. You can read straight through or dip in wherever your curiosity lands. Each chapter opens with a grounding scenario and closes with takeaways short enough to remember.

The first chapter covers what metagenomics actually is: how scientists moved from culturing bacteria on lab plates to reading the genetic signatures of entire microbial ecosystems. It is the methodological foundation for everything that follows.

The second chapter maps who lives inside you: the major bacterial groups, how composition shifts across your lifetime, and how the vaginal microbiome and pregnancy reshape communities. It also introduces the virome and mycobiome, the viral and fungal members that bacteria-focused coverage tends to leave out.

The third chapter asks whose "normal" we are measuring. Most of the reference data for microbiome testing comes from white, Western university populations, and the chapter explains what that means for people whose microbial communities do not match that baseline.

The fourth chapter explores the gut-brain axis, covering what is well-supported and, importantly, where mouse-study evidence and human clinical evidence have diverged.

The fifth chapter covers immunity. About seventy percent of immune tissue is located in and around the gut, and this chapter focuses on one of the most robustly supported areas of microbiome science.

The sixth chapter examines diet as the most modifiable factor shaping your microbiome, carefully distinguishing between studies that show microbial changes and those that show actual health outcomes.

The seventh chapter tackles probiotics (live microorganisms that, in adequate quantities, confer a health benefit on the host) and prebiotics (compounds that selectively feed beneficial microbes already resident in your gut), reviewing what the clinical evidence actually supports.

The eighth chapter covers the microbiome and disease, focusing on conditions where the link is strongest: C. diff infection, inflammatory bowel disease, and metabolic syndrome, as well as emerging evidence in colorectal cancer and long COVID.

The ninth chapter addresses fecal microbiota transplantation (FMT), now an FDA-approved treatment for recurrent C. diff infection, explaining how it works, what the safety data show, and why donor selection is complex.

The tenth chapter focuses on what you can actually do: how to interpret a stool report, which lifestyle changes have the most consistent support, and who stands to benefit from microbiome medicine first.

How to Read This Book

The straight-through path is the most complete, with each chapter building on the one before. But the book is also designed for readers who want to start at a specific point. Each chapter opens with a grounding scenario and closes with concise takeaways, which means any chapter can stand alone with enough context to be useful.

Wherever you see the evidence-tier labels, treat them as deliberate signals. Strongly Supported means you can rely on this with reasonable confidence. Promising means keep watching this space but do not reorganize your life around it yet. Marketing-Only means a claim is running ahead of its evidence and the right response is healthy skepticism. All key terms are defined at first use, and the glossary at the back collects them in one place. No prior biology knowledge is assumed.

Why This Book, Why Now

The Human Microbiome Project launched in 2007. In the years since, the peer-reviewed literature has grown faster than specialists can keep track of, while consumer testing companies have turned that momentum into a marketplace that has, predictably, outrun the data.

The person at that kitchen table deserves better than a wellness manifesto or an inaccessible academic review. They deserve an honest account of what the science has established, a clear explanation of how it works, and a practical framework for making decisions today while staying open to revision as new evidence arrives.

That is the book in your hands. Wherever you start, welcome.

1 The Hidden Universe

What Metagenomics Is and Why It Changes Everything

It is the summer of 1998, and Norman Pace is standing at the edge of a hot spring in Yellowstone National Park with a problem that would, in hindsight, reshape the entire discipline of biology. The water shimmers at temperatures approaching ninety degrees Celsius, yet the bottom of the pool blazes with color: vivid yellows, burnt oranges, deep greens, and brilliant reds, all produced by living microbial mats that carpet the superheated rock like alien paintwork. Something is clearly alive in there. The mats are thick, visibly organized, and metabolically furious, reproducing in conditions that would kill any animal within seconds. Pace and his colleagues know, from basic microscopy, that the mats are composed of microorganisms. But when they try to do what every microbiologist since Louis Pasteur has done as a matter of course, namely grow those organisms on agar plates in a laboratory, almost nothing survives. By every classical measure, the springs are almost sterile.

Pace refuses to accept this. He is not the first to notice the discrepancy between microscope counts and plate counts, a puzzle that has nagged at the field for decades under the dry technical label "the great plate anomaly." Still, he is among the first to propose a genuinely radical solution. Rather than keep trying to grow the organisms, what if you skipped that step entirely? What if you extracted all the DNA directly from the spring water, without ever trying to culture a single cell, and then read the genetic sequences to identify what was there? DNA does not need to survive in a Petri dish to tell you something. It just needs to exist. Pace's team developed the methods to do exactly this. When they applied them to the Yellowstone springs, they discovered communities of extraordinary richness and novelty: organisms no laboratory had ever seen, belonging to entirely new branches of the tree of life, thriving in a universe that science had, until that moment, lacked the tools to read.

Within a decade, Pace's insight catalyzed an entirely new field: metagenomics. Within two decades, that field was applied to the human body, producing the most detailed picture of the microbial communities living inside us that biology had ever assembled. Within three decades, the word "microbiome" was on the lips of wellness influencers, gastroenterologists, oncologists, and concerned parents in equal measure. But the story begins here, at the edge of a hot spring, with a scientist who refused to mistake the limits of his tools for the limits of the world.

Why It Matters

The story of Norman Pace at Yellowstone is not just a colorful anecdote from the history of science. It is the key that unlocks everything in this book. If you want to understand why the microbiome field exploded in the 2000s and 2010s, why thousands of research papers pour out of laboratories every year, why commercial stool testing is now a multi-hundred-million-dollar industry, you have to understand the single technological shift that made it possible: metagenomics, the ability to read all the DNA in an environmental sample without growing a single cell. Everything else- the discoveries, the controversies, the therapies, the hype- flows from that one change.

For more than a century, microbiology was, by necessity, a science of survivors. If a microbe could not be persuaded to grow on a laboratory plate, it was, for practical purposes, invisible. Researchers produced extraordinary results within those constraints, identifying the pathogens behind tuberculosis and cholera, developing antibiotics and vaccines that saved hundreds of millions of lives. But they were, without knowing it, studying a highly unrepresentative sample of microbial life. The organisms cultured in the lab were the outliers, not the majority. The majority was hiding in plain sight, in every gram of soil, every drop of seawater, every milliliter of human gut fluid, waiting for a technology that could read them without forcing them to survive conditions they had never evolved for. Metagenomics was that technology. And when it was finally turned on the human body, what it revealed was a universe.

The Great Microbial Blind Spot

To appreciate how profound the metagenomic revolution really was, you need to spend a moment inside the old world it replaced. Imagine being a microbiologist in, say, 1985. You have spent years mastering your craft: inoculating plates, maintaining sterile technique, coaxing finicky organisms into growing under the right conditions of temperature, atmosphere, and nutrient mix. You are, by any reasonable measure, excellent at your job. But the organisms you study are the easy ones. Such ones survive being yanked out of their natural habitat, transferred to an artificial growth medium, and incubated at a temperature convenient for a human laboratory. They are, in biological terms, the weeds, the generalists, the survivors. The species that depend on chemical signals from their neighbors, live in tight symbiosis with other organisms, and cannot function alone do not show up. They die on the plate or never grow at all, and so they are never cataloged, never named, never studied.

This is the great plate anomaly, and the numbers are genuinely astonishing. Scientists had long observed that microscope counts of microorganisms in soil or water samples typically ran into the many thousands of cells per milliliter. In contrast, plate counts of the same samples returned far fewer colonies, sometimes a hundred times fewer, sometimes a thousand times fewer. For a while, researchers assumed the discrepancy was a technical artifact that better technique would eventually resolve. But as the decades passed and the anomaly stubbornly persisted, a more unsettling conclusion took shape: somewhere between 99 and 99.9% of the microbial species in any given environment were simply beyond the reach of classical microbiology.

Diagram 1.1. Before and After: The Petri Dish Limit

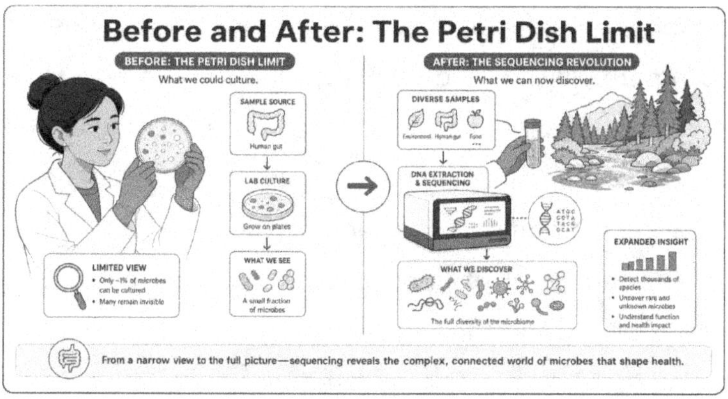

Think about what this means. For roughly a century, the entire edifice of classical microbiology rested on a method that, at best, captured one in a hundred of the species it was supposed to study. The textbooks, the reference databases, the classification systems, the medical diagnostics, all of it was built on a sample so biased it would make any epidemiologist wince. It is as if astronomers had cataloged the universe using a telescope that could detect only objects brighter than Venus, and then discovered that the vast majority of matter in the cosmos was invisible to their instrument. The picture they had assembled was radically incomplete.

Norm Pace was not alone in recognizing this. Carl Woese, whose work on ribosomal RNA in the 1970s had already revealed that the tree of life was far more complex than anyone suspected, had shown that a whole new domain of life, the Archaea, had been hiding under biologists' noses because it could not be cultured in ordinary conditions. But Pace took the next step, developing practical methods for extracting, amplifying, and sequencing environmental DNA when doing so required enormous technical ingenuity. The machinery was primitive by modern standards; sequencing a single gene fragment was slow, expensive, and error-prone. But the proof of concept was transformative. You did not need to grow the organism. You just needed its DNA. And its DNA was everywhere.

Diagram 1.2. The Great Plate Anomaly

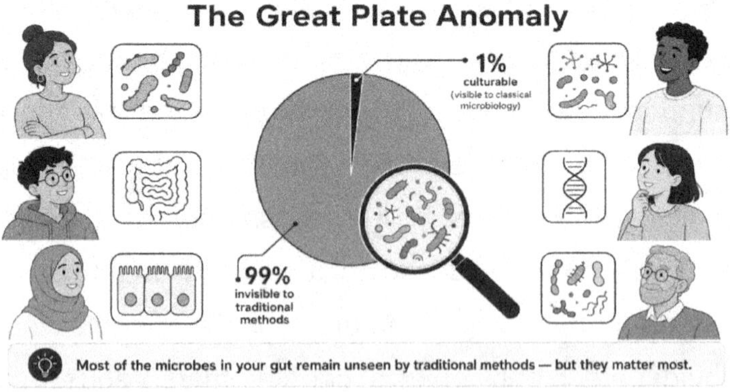

The implications spread outward slowly at first, and then very fast. If ninety-nine percent of microbial life had been invisible to classical methods, then every estimate of microbial diversity, every map of microbial communities, every conclusion drawn from culture-based studies would have needed to be revisited with humility and a new instrument. The human gut, long known to harbor bacteria, suddenly looked very different through this new lens. How many species were actually in there? What were they doing? How did they interact with each other, and with the human cells they were surrounded by? For the first time, biology had a way to ask those questions seriously.

What Metagenomics Actually Is

The word "metagenomics" sounds technical, but the concept is not complicated. The prefix "meta" comes from Greek and carries a sense of "beyond" or "transcending." A "genome" is the complete set of genetic information in an organism. Metagenomics is therefore genomics that goes beyond the single organism: it is the analysis of all the genetic material in a sample, regardless of which organism it came from. Rather than asking what the genes of this one bacterium are, metagenomics asks what all the genes present in this sample of soil, seawater, or human gut fluid are, from every organism present, simultaneously.

The practical process works roughly like this. A researcher collects a sample, say, a gram of human stool (a medium that turns out to be extraordinarily information-rich, since the gut microbiome sheds DNA constantly into the fecal stream). The sample is broken open using chemical and mechanical methods that lyse, or burst, all the cells present, releasing their DNA. That DNA is purified, fragmented into short fragments, and fed into a sequencing machine. Modern sequencing technology reads millions or billions of these short fragments in a single run, producing what is called a "shotgun" sequencing dataset, because, like pellets from a shotgun, the fragments cover the entire genetic content of the sample in overlapping, random pieces. Computational tools then reassemble these fragments, identify which organisms they came from by comparing them against reference databases, and quantify how many of each organism were present.

Diagram 1.3. The Library Shredder Analogy

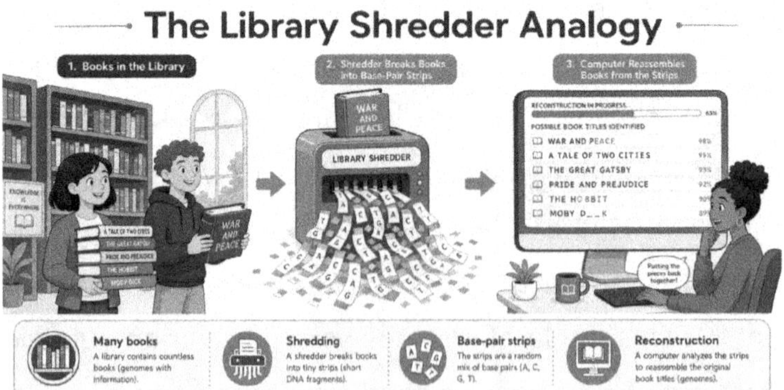

The library shredder analogy captures this well. Imagine a vast library containing tens of thousands of books, each representing the genome of a different organism. A metagenomics experiment is like feeding every book in that library through a shredder simultaneously and then reconstructing the titles and contents from the resulting pile of paper scraps. Each scrap is a short DNA sequence. The job of bioinformatics is to figure out which scrap came from which book, and to tell you not just what books were in the library but roughly how many copies of each were present. The "library" being shredded in a typical gut microbiome sample contains DNA from hundreds of bacterial species, dozens of viral strains, multiple fungal species, and human cells, all mixed in a single gram.

The analogy also clarifies why this is genuinely difficult. The shredded scraps do not come labeled. Many organisms share similar genetic sequences, just as many books share common phrases. The computational challenge of correctly attributing every fragment to its source organism remains an active area of research, and organisms not in the reference database are not identified. The method is not perfect. But imperfect and ninety-nine-percent complete is vastly preferable to perfect and one-percent complete, which is what the classical Petri dish offered.

It is also worth distinguishing metagenomics from a closely related technique called 16S rRNA sequencing, which uses a specific genetic marker to identify bacterial species in a sample. The 16S approach is cheaper and computationally simpler, and much of the early microbiome literature is based on it. But it only identifies bacteria, and only at roughly the species level. Shotgun metagenomic reads capture all the DNA in the sample, including viral, fungal, antibiotic-resistance, and metabolic-pathway genes, providing a much richer picture. Both approaches have their place; understanding the difference helps a reader navigate the scientific literature more critically.

The Scale of Microbial Life

Before turning to what metagenomics has revealed about the human body, it is worth pausing to appreciate the sheer scale of microbial life on Earth. Microorganisms are not a minor feature of life on Earth. They are the dominant form of life on Earth. They have been here for approximately 3.5 billion years, which is roughly seventy-five percent of the planet's entire history. Eukaryotes, the domain of life that includes all animals, plants, and fungi, are the late arrivals. We are guests in a microbial world that was already fully functioning when the first animal ancestor blinked into existence.

The numbers are genuinely difficult to hold in the mind. The observable universe contains an estimated two hundred billion trillion stars. The number of microbial cells on Earth is roughly ten to the power of thirty, or one nonillion, which is comparable to, and by some estimates exceeds, the number of stars in the observable universe. A single gram of healthy soil contains roughly one billion bacterial cells and ten million fungal cells. Your body, right now, contains something in the range of thirty-eight trillion bacterial cells, roughly comparable to the total number of human cells in your body. You are, by cell count, approximately half microbial.

Diagram 1.4. The Scale of Microbial Life

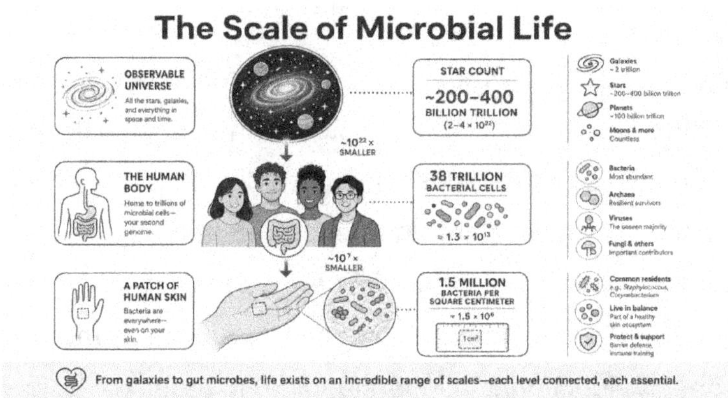

Microbes do not just outnumber everything else; they also outweigh most things. Bacteria fix atmospheric nitrogen into forms that plants can use, making all agriculture ultimately dependent on microbial chemistry. Certain marine microorganisms produce roughly half of all the oxygen in Earth's atmosphere, so that every second breath you take has been provided not by a rainforest but by microbes in the sunlit ocean surface.

They survive in conditions lethal to any animal. Thermophiles thrive in boiling water. Psychrophiles grow in Antarctic ice. Acidophiles prosper in mine drainage where the pH approaches zero. The organism Deinococcus radiodurans can survive a dose of radiation thousands of times greater than what would kill a human, reassembling its shattered genome with startling efficiency. Life, in the microbial world, is not a fragile enterprise. It is the most robust and persistent phenomenon this planet has ever produced.

Putting the human microbiome in this context changes how you think about it. The bacteria in your gut are representatives of the oldest lineage of life on Earth, living in a complex ecosystem that evolution has been fine-tuning for millions of years. The relationship between humans and our microbial inhabitants is ancient, mutual, and deeply functional. Understanding it required a tool commensurate with its complexity. Metagenomics provided that tool.

The Human Microbiome Project

In 2007, the National Institutes of Health announced one of the most ambitious coordinated research efforts in the history of biology: the Human Microbiome Project. The announcement came just a few years after the completion of the Human Genome Project, which had sequenced the entire human genome after more than a decade of international effort and at a cost of roughly $3 billion. The Human Microbiome Project was in some ways even more audacious, aiming to map the collective genomes of all the microbial organisms living in and on the human body, an enterprise so vast that no single institution could accomplish it alone.

The project united research centers across the United States and, in its second phase, internationally, around a shared ambition: to produce, for the first time, a comprehensive reference catalog of the healthy human microbiome. Researchers recruited 242 healthy adult volunteers and sampled them at 18 distinct body sites, including the gut, the mouth, the skin at multiple locations, the nasal passages, and, for female participants, the vaginal canal. Each site turned out to harbor a distinct microbial community, shaped by local chemistry, temperature, oxygen availability, and immune environment. The gut was the richest and most complex. But the findings from other sites were equally fascinating and, in some cases, more immediately clinically useful.

Diagram 1.5. The Human Microbiome Project at 18 Body Sites

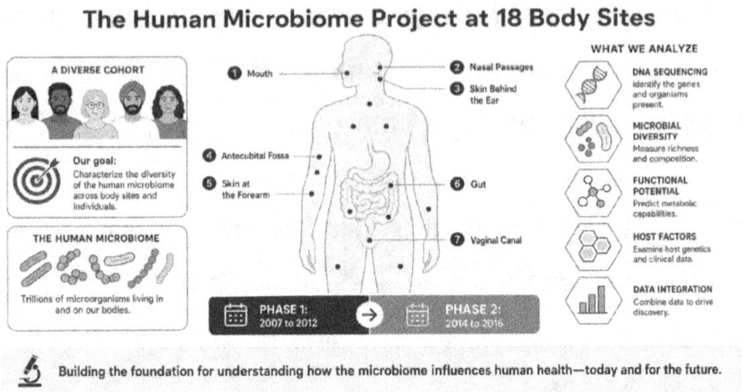

What the project found surprised nearly everyone, including the researchers who designed it. The first surprise was diversity. The variation in microbiome composition between healthy individuals was enormous. Two perfectly healthy adults could have gut microbiomes that looked almost nothing alike at the species level, yet both could be metabolically healthy and free of disease. This finding had a profound implication: there might not be a single "healthy microbiome" that everyone should aim for. Instead, health might be achievable through many different microbial configurations, much like a football team can win games with very different rosters as long as the core functions are performed. The concept of "functional redundancy" in the microbiome, the idea that many different species can perform the same critical biochemical job, became one of the field's most important frameworks.

The second surprise was stability. Despite the enormous variation between individuals, the microbiome within a given individual was remarkably stable over time. When the same participants were sampled months or even years apart, their microbial communities looked broadly similar, rebounding from perturbations such as antibiotic courses or illness to states close to their previous state. This resilience suggested that each person's microbiome was not a random collection of whatever happened to be passing through but a stable, organized ecosystem shaped by long-term host-microbe interactions, genetics, diet, and early-life exposures. Researchers began to describe it as a "personal core microbiome," a characteristic microbial signature that persisted over time, much like a person's fingerprints.

The third surprise was the sheer number of genes. When researchers added up all the unique genes in the human microbiome, they arrived at a number roughly one hundred times larger than the number of genes in the human genome itself. Your own DNA contains roughly twenty thousand protein-coding genes. The collective metagenome of your gut microbiota contains something in the range of two to four million. Those microbial genes encode enzymes that your own genome does not contain: enzymes for breaking down complex plant fibers your digestive system cannot process, enzymes for synthesizing vitamins your body depends on but cannot make, and enzymes for metabolizing drug compounds in ways that affect whether your medication works as expected.

The Human Microbiome Project was never designed to answer clinical questions. Its explicit goal was to establish a baseline reference, the equivalent of a detailed map before anyone had tried to navigate by it. But the map it produced was so rich, and the questions it raised so urgent, that it catalyzed an enormous wave of subsequent research that is still running at full force today.

Craig Venter and the Ocean Expedition

Before the Human Microbiome Project brought metagenomics to the human body, the technology had already proved its power in a more dramatic setting: the open ocean. In 2003, J. Craig Venter, the entrepreneur-scientist who had led Celera Genomics in the race to sequence the human genome, set sail on his research yacht, the Sorcerer II, to apply shotgun metagenomics to seawater at scale. The Sargasso Sea, a calm, relatively nutrient-poor stretch of the North Atlantic, was his first target.

Because it was nutrient-poor, classical microbiology had assumed the Sargasso Sea was a kind of microbial desert. If metagenomics could find surprising diversity even there, the implications for richer marine environments would be extraordinary. Venter's team collected water samples, extracted DNA, and sequenced it. The analysis revealed more than 1.2 million previously unknown genes and, depending on methodological assumptions, identified between 1,800 and 48,000 previously unknown microbial species in that stretch of ocean. Some turned out to encode entirely new forms of photoreceptor proteins, suggesting that the ocean harbored light-harvesting organisms previously unknown to biology.

Diagram 1.6. Craig Venter's Sargasso Sea Expedition

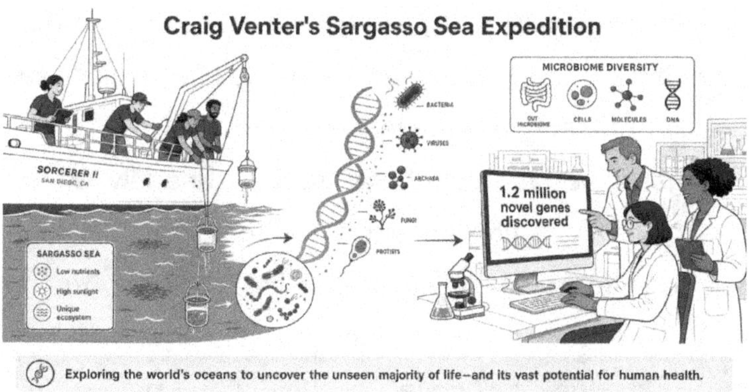

The Sorcerer II expedition did not stop with the Sargasso Sea. Over the next several years, Venter's team sailed an around-the-world route, collecting samples at dozens of ocean sites, and each site added thousands of new genes and hundreds of new species to the catalog. The cumulative dataset transformed our understanding of ocean biology, demonstrating that marine microbial diversity was orders of magnitude greater than classical methods had suggested.

For readers trying to grasp the power of the technology before it was applied to the human body, the Venter story is invaluable. Here is a person who sailed across the ocean, lowered buckets into the water, and came back with more than a million previously unknown genes from organisms no microbiologist had ever seen. The Yellowstone hot springs, the Sargasso Sea, and the human gut were all pointing toward the same conclusion: the microbial world is vastly larger and more diverse than classical biology had imagined, and metagenomics was the instrument that finally made it legible.

The Sargasso Sea dataset sat largely unanalyzed for years after its collection because the computational tools needed to interpret it did not yet exist at scale. Venter generated the data anyway. That is often how paradigm shifts work: the evidence accumulates before the framework is ready, and then the framework catches up in a rush.

Why This Changes Everything for Medicine

The connection between metagenomics and human medicine might seem, at first, like a long conceptual leap. What does reading DNA from seawater have to do with your digestive health, or your response to cancer treatment, or the effectiveness of your antidepressant? The answer is that metagenomics transformed medicine's ability to answer a question central to a startling range of clinical problems: exactly which microorganisms are present in this patient, and in what quantities?

Before metagenomics, clinical microbiology was, at its core, a culture-based discipline. When a patient came in with a suspected bacterial infection, the standard protocol was to culture a sample on agar for 24 to 48 hours and see what emerged. This worked reasonably well for common pathogens that are obligingly easy to culture. But it missed everything that would not grow on standard media, required days when hours might be critical, and told you nothing about antibiotic resistance genes lurking in organisms that happened not to grow on that particular plate.

Diagram 1.7. The Ripple Outward

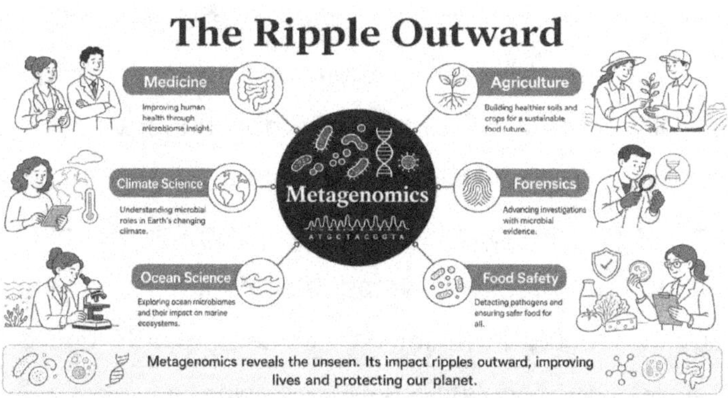

Shotgun metagenomics can identify every organism in a clinical sample in a single sequencing run that takes hours rather than days, without any culturing step. Researchers have used metagenomic sequencing to identify the causative agent in cases of mystery fever that resisted all classical diagnostic approaches; in one celebrated case, a child with encephalitis of unknown origin was diagnosed with a rare bacterial infection only after metagenomic sequencing of the cerebrospinal fluid revealed the pathogen. Hospitals are beginning to use metagenomics to track the movement of antibiotic-resistance genes through hospital wards in real time.

The ripples extend well beyond infectious disease diagnosis. One of the most striking recent findings is that a patient's gut microbiome composition at the time of cancer immunotherapy initiation significantly predicts whether the therapy will be effective. The immune system's responsiveness to checkpoint inhibitor activation is partly regulated by signals from gut bacteria, and patients with certain bacterial species in their microbiomes respond substantially better. Clinical trials are now underway to test whether modifying the microbiome before immunotherapy, through fecal transplantation or targeted probiotic interventions, can improve response rates. This is clinical oncology conducted in major cancer centers with rigorous trial designs, not speculative wellness science.

The same story is beginning to play out in psychiatry, cardiology, diabetes management, and pediatrics. Each chapter of this book explores one of these connections in depth. But the foundation for all of them is the same:

metagenomics made it possible to see the microbial communities involved in these processes clearly enough to study them systematically. The difference is between studying a forest by sniffing the air downwind and actually walking among the trees.

Why This Changes Nothing Yet

It would be a mistake, and this book will not make it, to translate the genuine excitement of the metagenomic revolution into a promise that medicine has already been transformed. The technology exists. The discoveries are real. But the gap between a research finding and a clinical intervention that reliably helps individual patients is large, and in the microbiome field, for most applications, it remains very wide.

The human gut microbiome contains hundreds of species engaged in thousands of simultaneous biochemical interactions, all of which interact with the host immune system, metabolism, nervous system, and genome. Identifying that "patients with more of species X respond better to treatment Y" is a correlation, not a mechanism. Understanding the mechanism well enough to design a reliable intervention requires a deeper level of knowledge that, for most of these associations, we do not yet have. The history of medicine is littered with enthusiastic correlations that did not survive the test of mechanistic understanding and rigorous clinical trial. There is no reason to assume the microbiome field is exempt.

Readers who have seen headlines announcing that a certain probiotic cures depression, or that fecal transplants can reverse obesity, or that the microbiome is the master key to all chronic disease, should be aware that these headlines have, nearly universally, outrun the evidence. The science is moving fast, but it is moving through a process that takes time, replication, and the hard work of distinguishing correlation from cause. The discoveries are real. The possibilities are genuine. The caution is equally necessary.

What This Book Is and Is Not

If so much is still uncertain, why should you read a whole book about it? The answer is that uncertainty, managed well, is more useful than false certainty. A reader who finishes this book knowing exactly where the science is solid, where it is promising, and where it is speculative, is far better equipped to make decisions about their own health than a reader who has been handed a simple story about good bacteria and bad bacteria and told to take a probiotic.

Consider a composite reader, drawn from the kinds of real people who seek out this subject. Meet Priya, a forty-two-year-old high school biology teacher in suburban Maryland. Priya knows more biology than most people, which makes her frustration with the microbiome literature especially acute. She received her commercial stool test report three weeks ago after a year of unpredictable digestive symptoms. The report arrived as a glossy booklet full of pie charts, bacterial names, percentages, and a "diversity score" comparing her to an anonymous reference population. She has read it twice. Her Bacteroidetes-to-Firmicutes ratio is flagged in amber. She does not know whether to worry, change her diet, order the recommended probiotic supplement, or call her gastroenterologist and read the report aloud over the phone.

Priya's confusion is not a failure of intelligence. It is the predictable result of being handed a document that reflects cutting-edge science without giving her the framework to interpret it. The stool test company is using real metagenomics technology, genuinely sequencing the DNA in her sample and identifying the species present. But the interpretive layer they have laid over those sequences- the traffic lights, the diversity scores, the supplement recommendations- sits far ahead of what the science currently supports for individual health guidance. Priya needs to know the difference between the sequencing result (real) and the clinical interpretation (often speculative). That is exactly what this book is for.

This book is a science communication project, not a wellness guide. It will not tell you which probiotic to buy. It will not tell you that your gut bacteria are making you depressed. It will not tell you that everything you think you know about human biology is wrong. What it will do is explain, as clearly and honestly as possible, what metagenomics is, what it has found, what the genuine uncertainties are, and how to think about all of this in a way that remains useful as the science continues to evolve.

Diagram 1.8. Excitement Plus Skepticism

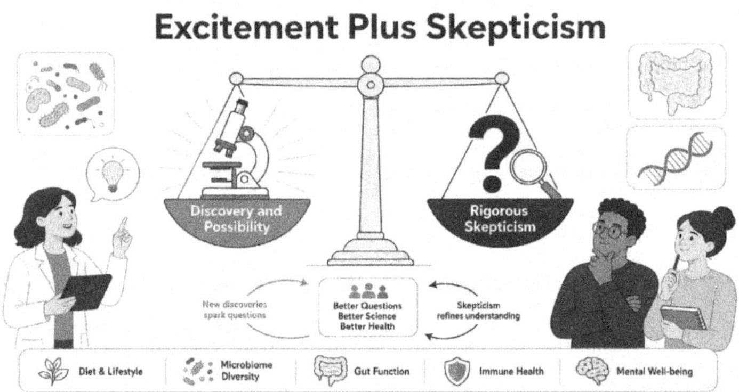

The book you are holding is not primarily a book about metagenomics as a technology. It is a book about what metagenomics has revealed, and what it has not yet revealed, about the relationship between the microbial world and human health. The technology is the key. The lock it opens is the human body. And what is behind the door is the rest of this book.

Priya does not need to understand every detail of shotgun sequencing to use this book well. She needs the conceptual framework: what the test is actually measuring; what "diversity" means and does not mean in a clinical context; what the Bacteroidetes-to-Firmicutes ratio does and does not predict; and what questions she should be asking her gastroenterologist rather than the ones the supplement company hopes she will ask. By the time she reaches the end of this book, she will have those tools. She will not have a magic solution to her digestive symptoms, because the science does not yet provide one reliably. But she will have something more durable: a framework for thinking about her microbiome that will remain valid as headlines change and the next wave of stool test companies arrives.

This is what good science education does. It does not hand you a conclusion. It gives you the lens. Once you have the lens, you can see for yourself.

Takeaways

- Metagenomics reads all the DNA in an environmental or clinical sample without growing any organisms, revealing the ninety-nine percent of microbial life that classical culture-based methods could not detect.
- The "great plate anomaly," the gap between microscope counts and plate counts, told biologists for

decades that something was wrong with their methods; metagenomics resolved it by bypassing the culturing step entirely.

- Norman Pace's work in Yellowstone, followed by Craig Venter's ocean expeditions in the 2000s, established that metagenomics could reveal enormous, unexpected diversity in any complex environment, from boiling springs to apparently simple ocean water.
- The NIH Human Microbiome Project (2007 to 2016) applied these methods to the human body for the first time at scale, documenting extraordinary diversity between healthy individuals and establishing that the human microbiome's collective genome contains roughly one hundred times more genes than the human genome itself.
- The medical implications are real and growing: metagenomic diagnostics, microbiome-informed cancer immunotherapy, and antibiotic-resistance surveillance are all moving from research into clinical practice.
- The gap between a research finding and a clinical intervention that reliably helps individual patients remains wide for most microbiome applications; correlation is not mechanism, and the field's genuine excitement should be held alongside genuine scientific rigor.

- The right posture for a reader of microbiome science is the same posture good scientists maintain: genuine curiosity, genuine excitement about what has been found, and a clear-eyed recognition that many of the most important questions remain, for now, open.

2 Your Personal Ecosystem

Consider a composite story drawn from many new parents who have found themselves deep in microbiome research late at night: Priya, a thirty-eight-year-old urban planner in Minneapolis, had not planned to become obsessed with bacteria. She ordered a home microbiome test for her four-month-old daughter, Leela, after reading about how birth mode shapes a baby's gut. Six weeks after mailing in the swab, a report landed in her inbox full of color-coded bars and Latin names she could not pronounce: Bifidobacterium longum, Lachnospiraceae, a "Diversity Score" hovering in the middle of a range from zero to a hundred. A note at the bottom said Leela's profile was "typical for her age." Priya stared at it for a long time, feeling something between relief and confusion. Typical compared to what? What did those names actually mean?

Leela is four months old when her gut is first documented, and that timing matters. The infant microbiome at four months is still in active assembly: dominated by Bifidobacterium in breastfed babies, still stratifying and settling, responsive to nearly every input. Whether Leela was born vaginally or by C-section, how long she was breastfed, whether she received antibiotics in her first weeks — each of these has already shaped the community in the report Priya is reading. The report cannot tell Priya what Leela's microbiome will look like at age five or fifteen. But it can serve as a baseline — a first photograph of an ecosystem that will change enormously before it stabilizes into something like an adult pattern around ages two to three. What Leela needs most right now is not optimization. It is time, diverse exposure, and continued breastfeeding if possible. The microbiome, at four months, is not something to fix. It is something being built.

She searched. She found dozens of articles using "microbiome" as if it were a single thing you could optimize. She found bloggers warning about antibiotics and pediatricians who dismissed the test entirely. None of it clicked until she stumbled across a sentence in a research review that stopped her scrolling: the human body, it said, is not one ecosystem but many, each as distinct from the others as a coral reef is from a mountain stream.

That sentence is a good place to begin. Before we talk about what your microbiome does -- before immunity, mental health, or metabolism -- we need to picture it clearly, neighborhood by neighborhood, noticing who lives there and why the species thriving on one block would be out of place three streets over.

Why This Tour Matters

Most of us grew up learning that bacteria make you sick. We washed our hands to remove them, took antibiotics to kill them, and bought antibacterial soap to keep surfaces free of them. That story is not entirely wrong, but it is radically incomplete. The same way that knowing "chemicals" make up everything you eat does not mean your meal is toxic, knowing that bacteria cover every surface of your body does not mean you are under constant attack.

What the last two decades of microbiome science have revealed is the opposite: a large, diverse, and well-functioning community of microorganisms is a defining feature of good health. Your microbiome is not the enemy. It is more like the infrastructure—the water mains and power lines that keep a city running invisibly until they fail. Getting a clear picture of who those organisms are and where they live is the foundation for everything else in this book.

Neighborhoods of the Body

When scientists say "your microbiome," they are not talking about a single community. They are talking about a collection of distinct microbial ecosystems, each shaped by conditions so different from one another that a species thriving in one location would die minutes after arriving in another.

Think of a well-run city with neighborhoods that look nothing alike: a harbor district, a dense downtown, a quiet suburb. Each neighborhood has its own population and its own rules. The human body is organized the same way.

Diagram 2.1. The Body as a City

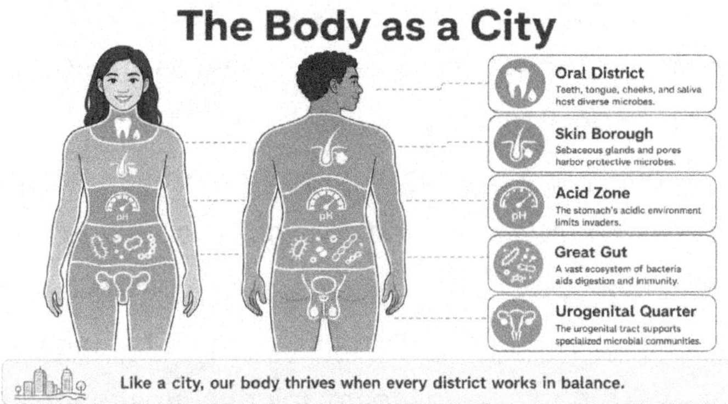

The mouth hosts more than 700 identified bacterial species. It is a warm, moist entry point constantly exposed to the outside world and washed in saliva rich in oxygen and sugars. Diets high in sugar favor acid-producing species that erode enamel. Diets rich in nitrate-containing vegetables favor species that convert those nitrates to nitric oxide, which benefits blood pressure.

The stomach, just centimeters south, is almost the opposite. Its pH can reach 1.5 to 2 -- roughly as acidic as battery acid -- and very few organisms survive there in large numbers. The most famous stomach resident, Helicobacter pylori, produces an enzyme that neutralizes its immediate environment. H. pylori is associated with ulcers and stomach cancer in some people. Yet, it lived in human stomachs for at least 100,000 years and was nearly universal in our species until recent decades. Its rapid disappearance from Western guts, driven partly by antibiotic use, may have unintended consequences. The story of H. pylori is an early lesson: a microorganism associated with disease in some contexts may have neutral or beneficial roles in others.

The small intestine has greater diversity than the stomach but remains modest compared to what lies downstream. Bile acids, digestive enzymes, and shorter transit times limit which species can establish there. When bacteria that normally live further downstream migrate upstream and proliferate, this results in small intestinal bacterial overgrowth (SIBO), which causes bloating, abdominal pain, and nutrient malabsorption.

The colon is the star of the microbiome show. It is warm, anaerobic (oxygen-free), rich in undigested plant fibers, and packed with 10 billion to 100 billion organisms per milliliter. This is where the most metabolically important microbial work happens: the production of short-chain fatty acids, the modulation of immune signaling, the synthesis of vitamins, and the degradation of compounds that your own enzymes cannot touch.

The skin hosts multiple ecosystems. Oily areas near the nose and ears favor lipid-loving Cutibacterium. Corynebacterium and Staphylococcus favor moist areas such as the armpits. Dry areas like the forearm have the lowest bacterial density. Your skin microbiome produces antimicrobial compounds and competes with pathogens for space and nutrients.

The urogenital tract has its own microbial character, especially in women -- a community distinctive enough to deserve its own section later in this chapter.

The 38 Trillion Number (and What It Means)

If the neighborhoods tour gave you a sense of the geography, the next question is about population: just how many organisms are we talking about?

The answer is 38 trillion -- the current best estimate of the bacteria living in and on the average adult human body, with the vast majority in the colon. The figure comes from a 2016 paper by Ron Sender, Shai Fuchs, and Ron Milo, who recalculated a number everyone had been repeating without checking the arithmetic. For decades, the standard claim was that microbial cells outnumbered human cells ten to one -- a rough 1972 estimate that was never rigorously verified. Sender and colleagues found something closer to one to one: roughly 30 trillion human cells, roughly 38 trillion bacterial cells, nearly equal.

These are not a handful of hitchhikers. They are a community so large it nearly matches the entire cellular count of a human body. To make it tangible: if the global human population were bacteria, you would need every person alive on Earth today multiplied by about 4,750 to match the count in a single human gut.

Diagram 2.2. The 38 Trillion Comparison

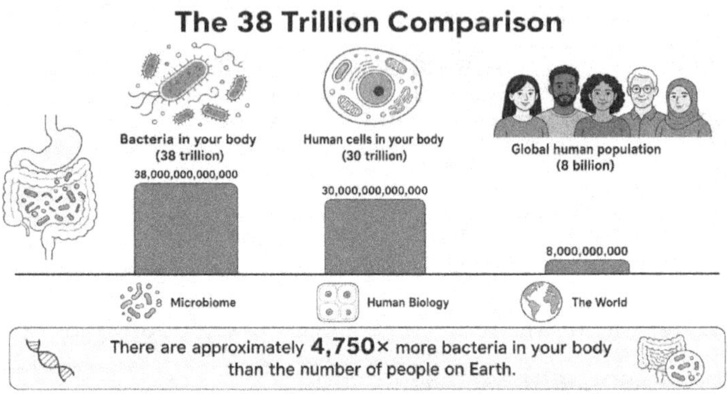

The raw number matters, but what matters more is what it implies about genetic capacity. Each bacterial cell carries its own genome. Add up the genetic content of all the microorganisms living in and on you, and you get the metagenome: the collective genome of the entire community. By some estimates, it contains roughly 150 times as many genes as the human genome itself. Many of those microbial genes encode proteins that do things your body cannot do on its own: break down plant fibers, synthesize vitamins, neutralize chemical compounds in food. You are borrowing their capabilities, and they are borrowing your warm, nutrient-rich interior. This is why the microbiome is not a footnote to human biology. It is a core operating system.

Those numbers set the stage for a more detailed look at who exactly is living inside the gut -- not just how many, but which organisms, what they do, and what happens when the balance among them shifts.

The Big Players (Firmicutes and Friends)

Despite the hundreds of species in a single healthy gut, the taxonomic landscape is not random. When microbiologists sequence gut communities from healthy adults, the vast majority of organisms fall into just a few major groups called phyla. Four dominate the human gut.

Diagram 2.3. The Big Four Phyla

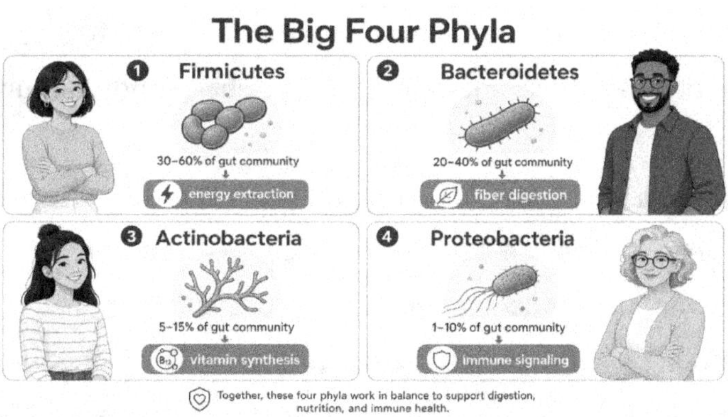

Firmicutes are the most abundant phylum in most healthy adults, making up 30 to 60 percent of a typical gut community. This huge and diverse group includes Lactobacillus (lactic acid producers found in yogurt and the small intestine), Clostridium (which includes both dangerous pathogens and beneficial butyrate producers), Ruminococcus, and Faecalibacterium. Firmicutes are specialists in energy extraction, particularly skilled at fermenting complex carbohydrates and producing short-chain fatty acids, especially butyrate -- the primary fuel for the cells lining your colon.

Bacteroidetes (now reclassified as Bacteroidota in some taxonomic systems, though the older name is still widely used) typically make up 20 to 40 percent of the gut community. Key genera include Bacteroides and Prevotella. Bacteroidetes are fiber digesters: they encode a remarkable variety of enzymes called carbohydrate-active enzymes (CAZymes) that break down the complex plant polysaccharides your own digestion cannot touch. The products of that fermentation -- acetate and propionate -- affect metabolism, appetite, and immune function.

Actinobacteria occupy 5 to 15 percent of the gut community. The most celebrated genus here is Bifidobacterium, which is abundant in breastfed infants and has been used in probiotic products for decades. Bifidobacteria specialize in digesting human milk oligosaccharides, the complex sugars in breast milk that an infant's own digestive system cannot break down.

Proteobacteria are present in small proportions (often less than 10 percent) in healthy guts but punch well above their weight in disease associations. This phylum includes E. coli (most strains harmless; a few dangerous), Salmonella,

Helicobacter, and Campylobacter. A small, stable Proteobacteria population is normal. An expanded one often signals dysbiosis—the clinical term for a microbial community that has shifted out of its healthy balance —and is frequently associated with inflammation.

Knowing a phylum's name does not tell you what it is doing in any particular person at any particular time. The microbiome is a living negotiation between organisms, incoming diet, immune signals, and dozens of other variables. Phyla are the broad strokes. The details lie in the species, strains, and metabolic outputs, which are where most of the interesting research is currently focused.

The Rainforest Principle: Diversity

A concept that will appear throughout this book is that diversity is a feature, not a coincidence.

In ecology, a community with many different species filling many different functional roles is more stable and more resilient than a simple one. A rainforest can withstand disturbances that would devastate a wheat field. The wheat field has one species doing one thing. Remove it, and the whole system collapses. The rainforest has a thousand redundant pathways. Knock out one species, and another fills its niche.

Diagram 2.4. Rainforest vs. Monoculture

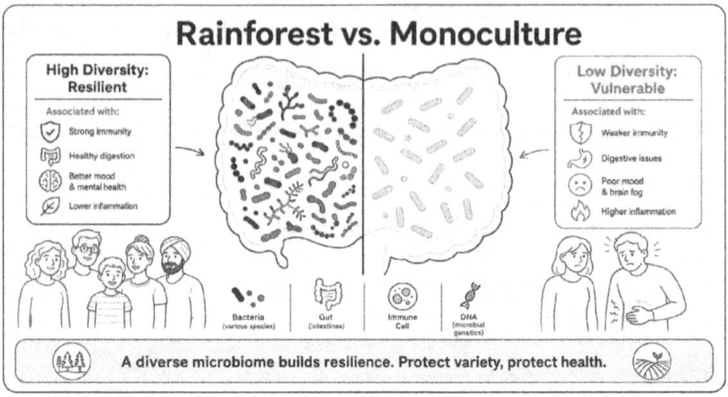

The human gut obeys the same logic. High microbial diversity -- measured by counting both the number of distinct species present (richness) and the evenness of their distribution (evenness) -- is one of the most consistent markers of gut health in the literature. Low diversity has been documented in people with obesity, type 2 diabetes, inflammatory bowel disease, colorectal cancer, rheumatoid arthritis, and asthma. The association is not perfect, but the pattern is robust.

Several mechanisms explain why diversity confers resilience. First, a diverse community is harder for a pathogen to invade because every ecological niche is already occupied—a principle known as colonization resistance. Second, a diverse community has metabolic redundancy: if one butyrate-producing species is temporarily depleted, another can step in. Third, a diverse community provides a broader range of signaling molecules to the immune system, helping calibrate immune tone—neither overreactive (allergies and autoimmunity) nor underreactive (vulnerability to infection). Diversity is a proxy for a healthy community, not the thing itself. A gut with hundreds of species, many of them inflammatory, would not be healthy just because it scores high on a diversity index. What matters is the right community, doing the right metabolic work. Diversity is a useful shorthand for that -- but still a shorthand. Most studies linking low diversity to disease are also observational: they cannot tell us whether low diversity caused the disease, whether the disease caused the low diversity, or whether both share a common driver, such as diet or antibiotic history. Causation in microbiome research is genuinely hard to establish, and this book will repeatedly be honest about that.

Your Core Microbiome

If your gut microbiome is as unique as your fingerprint, how can scientists say anything meaningful about what a "healthy" microbiome looks like? How can any general advice apply to everyone when everyone's starting point is different?

The answer lies in the core microbiome.

Despite enormous variation among individuals, researchers analyzing large cohorts of healthy people find that certain species and, more robustly, certain functional capabilities are present in the vast majority of healthy human guts. Not the same species at the same abundance, but the same functions reliably present: butyrate production, bile acid metabolism, vitamin synthesis, colonization resistance. Think of it as the difference between two neighborhoods that look nothing alike architecturally, but both have running water, electricity, and reliable mail delivery. The infrastructure is consistent even when the buildings differ.

Diagram 2.5. The Core Microbiome

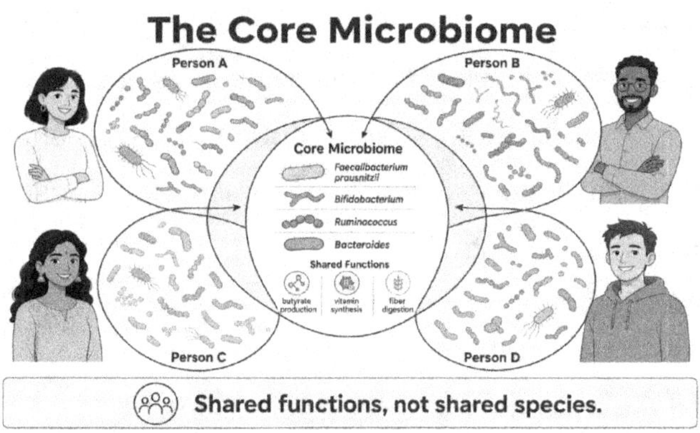

One well-studied core species is Faecalibacterium prausnitzii, a butyrate-producing Firmicute that is among the most abundant bacteria in a healthy adult colon and among the most consistently depleted in people with inflammatory bowel disease and metabolic syndrome. It is so strictly anaerobic that even a few seconds of oxygen exposure kills it -- which is one reason it took so long to study in a standard lab setting. Researchers now consider it a keystone species: one whose presence has an outsized stabilizing effect on the wider community.

More reliably than any specific species, the production of short-chain fatty acids (SCFAs) -- butyrate, propionate, and acetate -- appears in nearly all healthy human guts. These molecules feed the colonocytes lining the colon, dampen inflammation, influence appetite hormones, and communicate metabolic signals to the liver and brain. You can have very different species producing SCFAs in your gut versus mine, but as long as the production capacity is there, the function is maintained. This functional redundancy is what makes the core microbiome concept more powerful than any checklist of species.

The core concept also helps explain why microbiome testing that lists species is of limited practical value. A list of names without functional interpretation is like a city directory without a map. What matters is whether the functions are running -- which is why metagenomics -- sequencing not just which organisms are present but which genes they carry and which metabolic pathways they activate -- has become the gold standard in serious research.

The Microbiome's Weight and Its Origins

The collective wet weight of the bacteria in your colon is estimated at roughly 1.5 to 2 kilograms -- about the weight of the human brain. That figure surprises most people, because bacteria are individually microscopic. But 38 trillion of them packed together produce substantial mass. It is one of those numbers that makes the microbiome feel suddenly real: a pound and a half of living organisms in your colon right now, metabolizing, dividing, communicating, and producing compounds that cross the intestinal wall into your bloodstream.

Where did all of it come from?

The dominant view has long been that the womb is sterile and a fetus's gut is essentially empty. More recent studies have identified microbial DNA in placental tissue and meconium (the first stool a newborn passes), but many researchers argue those signals represent contamination or sequencing artifacts. The sterile-womb model remains contested and unresolved. What is not debated is what happens next.

Diagram 2.6. Origins of Your Microbiome

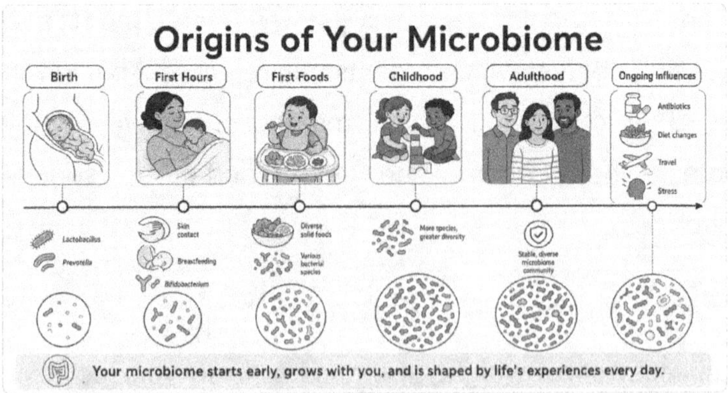

Birth is the most dramatic single microbial event in a human life. A baby passing through the vaginal canal is coated in its mother's vaginal and fecal microbiota -- a dense inoculation that colonizes the infant's gut, skin, and airway almost immediately. Vaginal births tend to seed infant guts with Lactobacillus and Prevotella, the same species that dominate the healthy vaginal microbiome. Babies born by cesarean section are first exposed to skin and environmental bacteria (Staphylococcus, Corynebacterium), and their early gut communities tend to resemble skin communities more than gut communities. Studies consistently show this early difference persists for at least the first year of life. Whether it has lasting effects on immune development is an active area of investigation, revisited in Chapter 5.

Breast milk adds another major layer of colonization. It contains complex sugars called human milk oligosaccharides (HMOs) that an infant's own digestive system cannot break down. These HMOs are not there to feed the baby -- they are there to feed specific bacteria, particularly Bifidobacterium longum subspecies infantis, which has co-evolved with humans specifically to metabolize them. Breastfed infants typically have gut communities dominated by Bifidobacterium, which produce an acidic environment that protects against harmful bacteria. When breastfeeding ends, and solid foods begin, the community diversifies rapidly. By about age three, most children's gut communities have taken on the general character of their adult counterparts -- though the assembly process continues for years, shaped by every meal, illness, and antibiotic course that follows.

Yours and Yet Changing

Two facts about the microbiome seem to be in tension until you examine them. First: your microbiome is uniquely yours. Second: your microbiome is constantly changing in response to what you eat, the medications you take, the stress you carry, and the people you live with. Both facts are true, and together they define the most interesting feature of the microbiome -- it is a personal ecosystem you are, in some measure, co-authoring.

The uniqueness was demonstrated by the American Gut Project, a crowd-sourced study organized by microbiologist Rob Knight involving tens of thousands of participants worldwide. Two people chosen at random typically share only 10 to 30 percent of their bacterial species. Even identical twins show surprising divergence over time. Genetics influences the microbiome, but environment, diet, and lifestyle contribute more to composition than shared DNA does.

The changeability has been documented in dozens of dietary intervention studies. Switch from an omnivorous diet to a plant-based diet for a few weeks, and measurable shifts appear in the gut community. Taking a course of broad-spectrum antibiotics can disrupt the community for weeks to months. Live with a partner long enough, and your gut communities become more similar to each other than either of yours would be to a stranger's -- a finding that researchers described, with admirable dry wit, as "microbial intimacy."

This changeability is a central source of hope. If your microbiome were fixed at birth and impossible to influence, there would be little point in studying how diet or interventions affect it. The fact that it changes means it can, within real limits, improve. Your microbial community is a living, responsive system, not a fixed fate.

The Women's Microbiome

The microbiome discussion in most popular books treats the human body as if it were generic: a standard-issue gut with standard-issue bacteria. That framing misses something important. The female body hosts microbial communities that are significantly shaped by hormonal cycles, reproductive events, and the unique biology of the vaginal environment. For roughly half the population, these dynamics are not a footnote to microbiome science -- they are a central part of it.

The Vaginal Microbiome

The vaginal microbiome occupies a peculiar and fascinating position: it is one of the few body sites where low diversity is considered healthy. While a diverse gut is associated with resilience, the healthy vaginal microbiome in reproductive-age women is typically dominated by one or a few Lactobacillus species, particularly Lactobacillus crispatus, Lactobacillus iners, Lactobacillus gasseri, and Lactobacillus jensenii. This dominance of Lactobacillus is not a sign of a depleted ecosystem. It is an evolved protective state.

Lactobacillus species produce lactic acid, keeping the vaginal pH low—typically between 3.8 and 4.5, about as acidic as tomato juice. This acidic environment is hostile to most pathogens, including the bacteria associated with bacterial vaginosis (BV), sexually transmitted infections, and some fungal overgrowths. A vaginal microbiome dominated by L. crispatus is associated with the lowest risk of BV and the best vaginal health outcomes in the research literature. L. iners is more ambivalent -- it tolerates a wider pH range and tends to persist even in disrupted communities, making it less reliably protective.

Not all women have Lactobacillus-dominant vaginal microbiomes. Research conducted in the United States shows that Black and Hispanic women are more likely than white women to have diverse, polymicrobial vaginal communities with lower Lactobacillus abundance. This finding has generated significant scientific and ethical debate, because much of the reference data to date has come from predominantly white, Western populations, and defining "healthy" based on those populations risks pathologizing normal variation in others. This tension in the field will recur in Chapter 3.

Diagram 2.7. The Women's Microbiome Across Life Stages

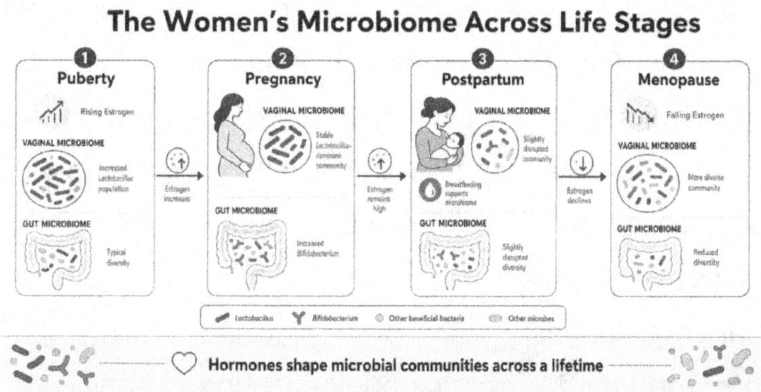

Pregnancy and the Shifting Microbiome

Pregnancy produces some of the most dramatic and well-documented microbiome changes in human biology. The vaginal microbiome typically shifts during pregnancy toward greater Lactobacillus dominance and reduced diversity—the opposite of what happens in the gut. Researchers believe this is a protective adaptation: a more stable, acid-producing vaginal community reduces the risk of ascending infections that could threaten the pregnancy.

The gut microbiome also changes substantially across the three trimesters. Studies have found that by the third trimester, the gut microbiome looks quite different from that of early pregnancy, with increases in Firmicutes and decreases in diversity—changes that, in non-pregnant individuals, would be associated with metabolic dysfunction. In pregnancy, these shifts appear to be adaptive, supporting the caloric demands of fetal growth. Immune function is recalibrated during pregnancy, with a shift toward less inflammatory immune tone (to avoid rejecting the fetus), which has downstream effects on which gut microbes can thrive. The physical expansion of the uterus also slows gut transit time, changing the colon's microbial conditions -- one reason many pregnant women experience constipation.

Postpartum Changes

The postpartum period brings another round of transitions. The vaginal microbiome, which was relatively stable and Lactobacillus-dominant during pregnancy, tends to shift again after delivery—often transiently toward lower Lactobacillus abundance before stabilizing. Women who deliver vaginally typically re-establish their pre-pregnancy vaginal microbiome within weeks to months. Women who deliver by cesarean may show a slower return to baseline.

For the gut microbiome, the postpartum period is complicated by overlapping factors: hormonal changes, sleep disruption, the caloric demands of breastfeeding (if applicable), and potential antibiotic exposure during delivery. This is where Priya's story connects directly to

Leela's test results. The infant microbiome is shaped by maternal factors: the bacteria colonizing the birth canal, whether delivery was vaginal or cesarean, whether breastfeeding followed, and the mother's gut community during pregnancy. These are not trivial influences. They are the starting conditions of a lifelong microbial community.

Menopause and the Aging Female Gut

As estrogen levels fall at menopause, the vaginal microbiome typically loses its Lactobacillus dominance and shifts toward higher diversity. This is associated with thinning and drying of vaginal tissue (vaginal atrophy) and increased risk of urinary tract infections and vaginal discomfort. Estrogen is critical for maintaining the glycogen-rich vaginal environment that Lactobacillus species depend on for fuel. Without glycogen, Lactobacillus populations decline.

The gut microbiome also changes at menopause. Studies comparing pre- and postmenopausal women find shifts in Firmicutes-to-Bacteroidetes ratios and changes in diversity that parallel the metabolic changes many women experience at midlife: shifts in body fat distribution, changes in insulin sensitivity, and altered inflammatory tone. Whether the microbiome drives these metabolic shifts, responds to them, or co-varies with them remains under investigation. One established link: estrogens are partially metabolized in the gut by a community of bacteria called the estrobolome, and disruptions to the estrobolome may affect circulating estrogen levels. For women, the microbiome is a dynamic participant in the hormonal story of a lifetime—not a passive bystander.

Beyond Bacteria: The Virome and Mycobiome

Everything described so far has focused on bacteria -- with good reason, since bacteria are the most numerous and most studied members of the microbial community. But the human microbiome is not a bacterial monoculture. Two other groups of microorganisms share our bodies and deserve a brief introduction before we close our neighborhood tour.

Diagram 2.8. The Hidden Members: Virome and Mycobiome

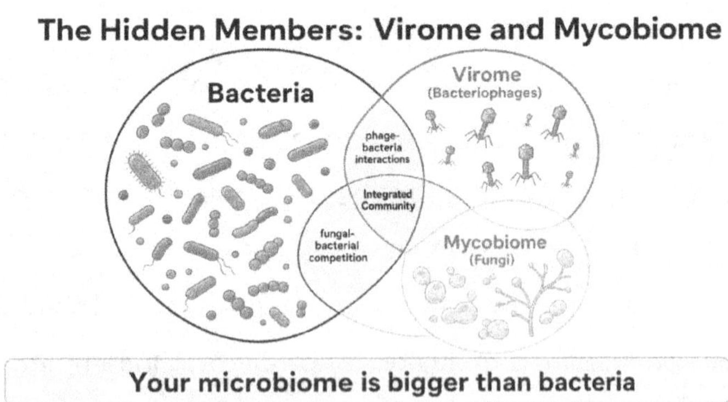

The virome is the collection of all viruses associated with the human body. Most of these are bacteriophages (phages): viruses that infect bacteria rather than human cells. Phages are, by some estimates, the most numerous biological entities on the planet. In the gut, they appear to play a key role in shaping bacterial community structure, selectively killing certain species and maintaining a predator-prey dynamic that influences which bacteria dominate at any given time. The gut virome is poorly characterized compared to the bacterial microbiome -- standard DNA sequencing methods are optimized for bacterial genomes and often miss viral ones -- but what research exists suggests it is as individually variable as the bacterial community and as responsive to diet and disease.

The mycobiome is the fungal component. Fungi are present throughout the body, most notably in the gut and on the skin. Candida species are the most widely studied gut fungi, present at low levels in most healthy individuals and capable of overgrowth when the bacterial community is disrupted—for instance, after antibiotic treatment. Saccharomyces, Aspergillus, and other genera contribute to the gut fungal community. Research on the mycobiome lags behind the bacterial field. Still, early work suggests that gut fungi communicate with the immune system and interact with bacteria in ways that may be relevant to inflammatory bowel disease and other conditions. When scientists talk about restoring a disrupted microbiome, they usually think of bacteria—but a complete picture includes fungal and viral communities as well.

That earlier introduction to the mycobiome matters here because antibiotic disruption is never only a bacterial story. When antibiotics clear ecological space in the gut, fungi such as Candida can expand into that opening, which is one reason a complete recovery story has to include the fungal community rather than treating the microbiome as bacteria alone.

Takeaways

- The human body hosts many distinct microbial ecosystems rather than one: the acid-tolerant stomach community, the oxygen-free fiber-fermenting colon, the Lactobacillus-dominant vaginal environment, and the skin's geography of oily, moist, and dry zones are all radically different neighborhoods.
- Roughly 38 trillion bacteria live in and on the average adult (concentrated mostly in the colon), with a collective genome encoding approximately 150 times the gene count of the human genome -- a massive reservoir of borrowed biological capability.
- The gut community is dominated by four phyla: Firmicutes, Bacteroidetes, Actinobacteria, and Proteobacteria, each contributing distinct metabolic

functions; high Proteobacteria relative abundance is a common marker of dysbiosis.

- High microbial diversity is a reliable marker of a resilient gut, providing colonization resistance, metabolic redundancy, and immune calibration; low diversity is consistently associated with chronic disease, though the causal direction is still being actively worked out.
- A shared set of core functions (butyrate production, fiber digestion, vitamin synthesis) appears across most healthy human guts despite enormous species-level variation between individuals -- function matters more than any checklist of species names.
- The female microbiome is shaped by estrogen and progesterone across a lifetime: the vaginal microbiome is healthiest when Lactobacillus-dominant, and both vaginal and gut communities shift significantly during pregnancy, postpartum, and menopause.
- The microbiome includes more than bacteria: bacteriophages in the virome and fungi in the mycobiome are integral members of the community, increasingly recognized as participants in health, disease, and the microbial ecology of the gut.

3 Reading the Code

Imagine receiving a 500,000-piece jigsaw puzzle with no picture on the box. Half the pieces are nearly identical in color, thousands are missing entirely, and a few dozen are from a completely different puzzle that ended up in the same bag. That is roughly the challenge a bioinformatics team faces when they open a raw metagenomic sequencing file. The raw data is not a tidy list of species names. It is a torrent of genetic text fragments, billions of short strings of the letters A, T, C, and G, with no obvious order. And yet, within hours, specialized software can piece together who was in that sample, what those organisms were likely doing, and how their microbial community compares to thousands of reference samples from around the world.

That process, from stool sample to colorful bar chart, involves some of the most remarkable technology in modern biology. You do not need a biology degree or any prior exposure to genetics to follow it. You need only the willingness to trace the thread from a cotton swab to a computer screen, and the curiosity to ask: what is this technology actually telling me, and what is it leaving out?

Microbiome testing is no longer confined to research labs. For under two hundred dollars, you can mail a sample to a consumer testing company and receive a personalized report on your gut's microbial landscape. Millions of people have done exactly that. Understanding how those reports are generated, what the science can and cannot support, and why results from the same sample can look different at two different labs gives you the tools to read any microbiome report or research headline with clarity and healthy skepticism.

Why This Chapter Matters

There is a gap in popular science coverage of the microbiome. Journalists and wellness brands tend to skip from "scientists collected stool samples" to "scientists discovered that X bacteria is linked to Y condition." The middle part, how a biological sample becomes interpretable data, gets compressed into a vague gesture toward "advanced testing." That compression hides almost everything you need to evaluate a claim.

Understanding the two main sequencing methods, knowing why bioinformatics pipelines depend on reference databases built by fallible humans, and grasping why a test kit report carries meaningful uncertainty does not diminish the science. It makes you a smarter participant in it. When you finish this chapter, you will understand why the same sample sent to two labs can produce different reports, why most consumer test results cannot predict disease, and why roughly half the microbial species in your gut may still be unknown to science.

DNA Sequencing: The Rosetta Stone of Biology

To understand microbiome testing, you first need a feel for what DNA actually is and why reading it unlocks so much biological information.

Every living cell stores its instructions in DNA. DNA is a long molecule built from four chemical units called nucleotides, each defined by a nitrogen-containing base. The four bases are adenine (A), thymine (T), cytosine (C), and guanine (G). That is the entire alphabet of life. Every trait an organism possesses, every protein it can build, every behavior it is capable of, is encoded somewhere in the precise sequence of those four letters.

In a human cell, the full DNA sequence is about 3.2 billion letters long. A typical gut bacterium carries a genome of 2 to 10 million letters. A single gram of human stool contains the DNA of hundreds or even thousands of different microbial species, all mixed in a biological soup that also includes shed human intestinal cells and food residues. The challenge of microbiome science is accurately interpreting that mixture and making sense of what is there.

Diagram 3.1. The DNA Alphabet

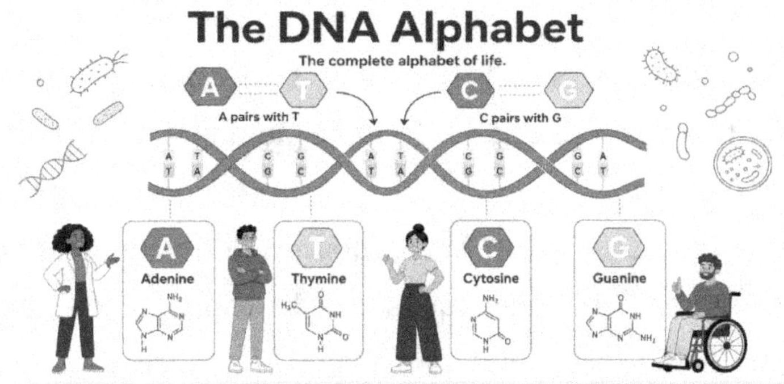

Diagram 3.1. The DNA Alphabet.

DNA sequencing is the process of reading the order of those letters in a sample. For most of the twentieth century, it was slow, expensive, and reserved for short stretches of DNA. The Human Genome Project produced the first complete human DNA sequence only after more than a decade of international effort and nearly three billion dollars. Today, a modern sequencing machine can read hundreds of millions of DNA fragments in a single run, and the cost has dropped to the point that sequencing an entire bacterial genome costs a few dollars.

This revolution, called next-generation sequencing, is what made microbiome science possible at scale. Instead of studying one organism at a time in a culture, scientists could take a sample directly from the environment and sequence everything in it at once. This approach, studying genetic material drawn directly from an environment without growing anything in a culture dish, is called metagenomics. The name is fitting: metagenomics goes beyond the limits of traditional microbiology, which could only study organisms willing to grow on laboratory plates. Many gut bacteria refuse to do that. Metagenomics reads their DNA directly.

A useful analogy is a city's phone directory. Before metagenomics, microbiologists could only study the residents willing to grow in a lab. Sequencing is like receiving the complete directory: every name, every address, and if the listings are detailed enough, every job title. The directory is not a perfect substitute for walking the streets, but it gives you an overview that was simply impossible to construct before.

16S rRNA: The Microbial Barcode

If sequencing all the DNA in a sample is like reading the entire phone directory, there is a faster, cheaper alternative that focuses on just one field in every entry. Every bacterium on Earth carries a gene called 16S ribosomal RNA (16S rRNA). This gene encodes part of the ribosome, the cellular machine all living cells use to build proteins. Because every bacterium needs ribosomes to survive, the 16S rRNA gene is present in every bacterial cell without exception.

What makes this gene so useful is its structure. Some regions are nearly identical across all bacteria ("conserved" regions) and serve as anchor points for the laboratory tools called primers that start the sequencing process. Other regions vary from species to species ("variable" regions) and hold the species-identifying information.

The result is a universal barcode. By targeting just the 16S rRNA gene from every organism in a sample, researchers can profile the bacterial community without sequencing all of the DNA. The sequencer reads the variable regions of thousands of 16S gene copies, and those sequences are compared to large reference databases to assign species or genus-level identities. The process is like checking barcodes on grocery items without unpacking any boxes. You get a reliable inventory of what is there, but you learn nothing about the contents.

Diagram 3.2. The 16S rRNA Barcode

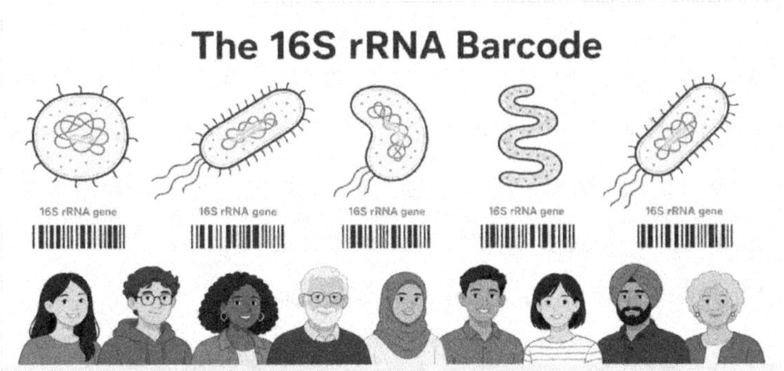

Diagram 3.2. The 16S rRNA Barcode. Same gene, different identity.

This targeted approach dominates the consumer microbiome testing market for practical reasons. A single sequencing run using this method costs far less than whole-genome approaches, the data files are smaller, and the interpretation pipeline is well established. For a company processing thousands of samples per day, these advantages are significant.

The limitations are equally important. First, 16S rRNA sequencing identifies bacteria but is not reliable for identifying fungi, viruses, or other members of other microbial kingdoms. Second, the resolution is often limited to the genus level, indicating that a certain Lactobacillus is present but not which species, and species within the same genus can have very different effects in the gut. Third, knowing a species is there tells you nothing about whether its genes are actively expressed. A bacterium can be present without being particularly active. For all its efficiency, 16S rRNA sequencing gives you a census, not a live feed.

A census is genuinely valuable. This method has produced large population-level datasets that underpin much of what we know about how microbiome composition shifts with age, diet, antibiotics, and disease. It is the workhorse of microbiome epidemiology.

Shotgun Metagenomics: The Full Picture

Shotgun metagenomics removes the targeting step entirely. Instead of hunting for one specific gene, it sequences everything in the sample: human intestinal cells, bacteria, fungi, viruses, bacteriophages, archaea, and entirely unknown entities. The result is a data tsunami:

billions of short DNA fragments drawn from every organism that contributed genetic material to the sample.

Rather than using a precision instrument aimed at a single target, you load the sequencer and blast the entire sample. What you lose in precision you gain in breadth. Shotgun data can tell you not only who is present but what those organisms are genetically equipped to do. Because you are reading genes across entire microbial genomes, you can ask: does this sample contain antibiotic resistance genes? What metabolic pathways are most active here? Does this sample contain organisms whose DNA matches nothing in any reference database?

Diagram 3.3. Shotgun vs Targeted Sequencing

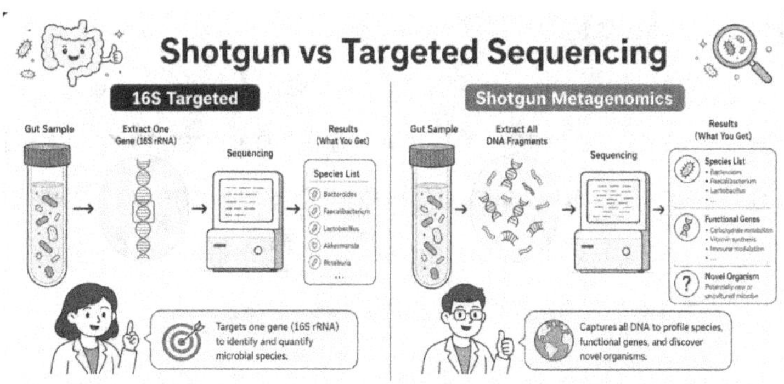

Diagram 3.3. Shotgun vs Targeted Sequencing

Shotgun metagenomics is increasingly used in clinical research to characterize disease-associated microbiomes, identify probiotic targets, and discover entirely new bacterial species. Some of the most influential microbiome science of the last decade, including studies linking gut microbiome profiles to immunotherapy outcomes in cancer patients, relied on shotgun data because it needed functional gene information that 16S sequencing cannot provide.

The trade-offs are substantial. Shotgun metagenomics is considerably more expensive than 16S sequencing, often by a factor of five to ten, and the data files are dramatically larger. Most consumer testing companies still rely on 16S or a related targeted approach because shotgun metagenomics at consumer scale is not yet economically practical. One exception is Viome, which uses a form of RNA sequencing to capture gene expression activity rather than just gene presence. We will return to a case story involving Viome later in this chapter.

There is also the challenge of human DNA. In a stool sample, a large portion of the DNA comes from shed intestinal epithelial cells. Shotgun sequencing reads everything, so the pipeline must first identify and discard human DNA before microbial analysis can begin.

Bioinformatics: Making Sense of the Data

A typical shotgun metagenomics run generates between one and ten billion short reads, each between 150 and 300 letters of genetic text. The sequencer produces this as a raw text file: billions of short excerpts from the genomes of thousands of organisms, all mixed without labels. Turning that raw text into a meaningful picture requires a sophisticated sequence of computational steps known as a bioinformatics pipeline.

Think of a factory assembly line where raw material enters at one end and a finished product exits at the other, with each station performing a distinct transformation. Raw sequence reads enter at the beginning and pass through quality filtering (removing reads that are too short or too error-prone), alignment (matching each read to sequences in a reference database), assembly (stitching overlapping reads back into longer genome stretches when possible), taxonomic classification (assigning species or genus labels based on closest database matches), and finally statistical analysis and visualization, which produces the charts that researchers and consumers actually look at.

Diagram 3.4. The Bioinformatics Pipeline

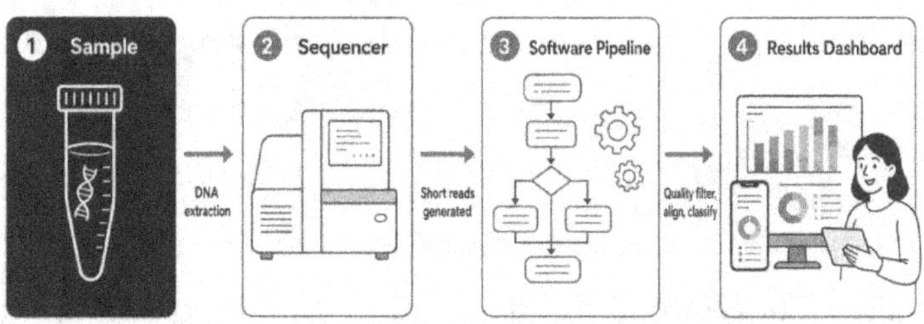

The Bioinformatics Pipeline

Each station in the pipeline can introduce variability. The DNA extraction method matters enormously: different kits break open bacterial cell walls with varying efficiency so that some species may be systematically over- or underrepresented. The sequencing platform affects which reads survive quality filtering. The reference database determines which organisms can be identified at all.

The reference database deserves particular attention because it is where much of the interpretive power and much of the uncertainty live. The most widely used databases for 16S rRNA classification include SILVA and Greengenes, both cataloging bacterial sequences from thousands of studies over decades. These databases are updated as new sequences are submitted, and a species unidentifiable two years ago may be identifiable today. Results from an older analysis and a newer one of the same sample might differ simply because the database changed, not because the biology did.

For a reader evaluating a consumer test kit, the pipeline is mostly invisible. The company runs its own proprietary version of these steps and uses its own reference database version. This is not inherently dishonest, but it does mean that comparing results from Viome and Ombre is not straightforward, even if you sent the same sample to both. They are running different pipelines, and the results reflect those differences as much as they reflect the biology of your gut.

Consumer Microbiome Testing: What the Kit Measures

Consider a composite story drawn from many real experiences. Marcus is fifty-three years old, college-educated, and has spent two years reading about the gut microbiome in books, podcasts, and magazine articles. His digestive health is fine by most measures, but he has noticed some bloating after certain meals and mediocre sleep. He spends $179 on a Viome Gut Intelligence Test, collects a stool sample using the provided swab and stabilization kit, seals it in the prepaid mailer, and drops it at a shipping location on a Tuesday. Six weeks later, his results are ready in the Viome app.

What Marcus is about to read is the output of the process described in this chapter. A laboratory technician extracted RNA from his sample (Viome uses metatranscriptomic sequencing, which reads messenger RNA rather than DNA and captures which genes are actively being expressed), ran the material through the company's sequencing platform and proprietary bioinformatics pipeline, compared his results to the company's internal reference population, and generated a dashboard describing his "gut microbiome age," a list of organisms and their relative abundances, and personalized dietary recommendations labeled "Superfoods," "Enjoy," "Minimize," and "Avoid."

Diagram 3.5. Inside a Consumer Test Kit

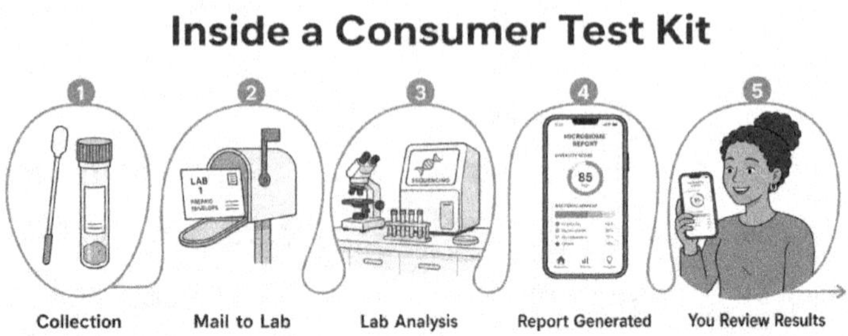

Diagram 3.5. Inside a Consumer Test Kit.

Marcus's experience will be familiar to anyone who has used one of these services. The app is well-designed and written in a friendly language. The list of organisms includes names like Akkermansia muciniphila (flagged as beneficial), Faecalibacterium prausnitzii (also described positively), and several Bacteroides species. The report has the confident, personalized feel of a medical document.

What Marcus is reading, however, is a combination of two very different things. The first is a genuine biological measurement: RNA was extracted from his sample, sequenced, and processed through a pipeline that yielded real data on which microbial genes were expressed at the time of collection. The second is an interpretive framework: Viome's algorithms translate those measurements into health scores and dietary advice using correlations derived from the company's reference population. Whether those correlations are predictive enough to guide individual dietary choices remains a scientific question.

Responsible use of a consumer test kit means holding both facts at once. The technology is measuring something real. The interpretation of what that measurement means for your individual health is considerably less certain. Marcus's "Avoid" list includes chia seeds, which had been a regular breakfast ingredient for him for three years with no apparent ill effect. A thoughtful reading treats that as a hypothesis to consider, not an instruction to follow.

The consumer testing landscape includes several companies worth knowing. Viome uses metatranscriptomic sequencing and focuses on functional gene expression. Ombre uses 16S rRNA sequencing and emphasizes probiotic recommendations. Zoe combines microbiome analysis with blood sugar monitoring and has published peer-reviewed research on its methodology. Each company uses a different technology, a different reference population, and a different interpretive framework. For a consumer trying to compare results across platforms, this poses a fundamental challenge: the reports do not speak the same language, even when they appear to.

Accuracy, Reproducibility, and the Limits

One of the most important, and underreported, stories in microbiome science is what happens when researchers test the reproducibility of these results. Reproducibility is a core scientific value: a good measurement should produce similar results when repeated. For microbiome testing, reproducibility is a genuine challenge at multiple levels.

The first challenge is biological. Your microbiome is not a fixed entity. It shifts daily in response to what you ate yesterday, whether you slept poorly, and what antibiotics you took months ago. Two samples from the same person taken one week apart can show meaningful differences in the relative abundances of specific taxa. This means that even a perfectly accurate test would show different results at different time points, not because the test is wrong but because the biology changed.

The second challenge is technical. In a landmark study conducted around 2019, researchers sent the same biological sample to several different microbiome testing labs and compared the results. The study found substantial variability across labs. Some variability came from differences in DNA extraction methods. Some came from which reference database each lab used. Some came from bioinformatics pipeline parameters. The overall picture: impressive technology does not automatically produce standardized, reproducible results sufficient for confident individual-level clinical predictions.

Diagram 3.6. Same Sample, Different Results

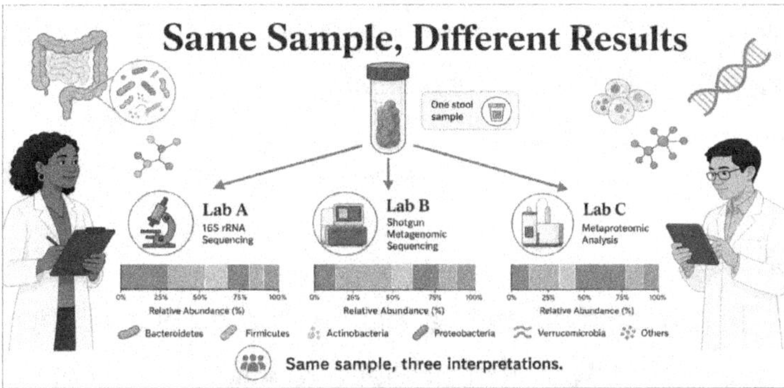

Diagram 3.6. Same Sample, Different Results

This does not mean the technology is useless. It is in an adolescent phase, powerful and promising but not yet mature enough for the precision medicine that some marketing materials suggest. Population-level research has yielded valuable insights into how diet, geography, age, and disease relate to microbial community structure. The leap from "this community pattern correlates with this condition in a study of ten thousand people" to "your specific profile predicts you will develop this condition" is still, in most cases, a leap the data cannot fully support.

Other sources of variability include the time between collection and processing (stool sitting at room temperature too long can shift microbial composition), the specific storage buffer in consumer kit tubes, sequencing depth (a shallow run will miss rare species that a deeper run would detect), and batch effects (systematic differences when samples are processed at different times or on different machines). Consumer testing companies handle these variables with varying degrees of rigor, and most do not make their validation data readily available.

What a Read Actually Is, and Why Missing Pages Matter

You now understand that a sequencer produces short fragments of DNA text called reads, and that a bioinformatics pipeline matches those reads to reference databases. There is a layer beneath this that is worth making explicit, because it carries profound implications for everything the field can and cannot claim.

A single sequencing read is typically between 150 and 300 letters of genetic text. On its own, it means nothing to a human reader. The bioinformatics software asks: where in the collection of known genomes does this sequence appear? If it matches a known stretch of Bacteroides fragilis genome, the read gets that identity. If it matches multiple species, it is assigned to a broader taxonomic group. If it matches nothing, the read is classified as unassigned, essentially invisible to the analysis.

The book-shredding analogy captures this vividly. Imagine every genome in your gut microbiome is a book in a vast library. Shotgun metagenomics shreds every book simultaneously into confetti-sized pieces. The bioinformatics pipeline attempts to reassemble the confetti by comparing each piece to a master catalog of known books. For well-studied bacteria whose full genome sequences have been published and deposited in databases, reassembly works reasonably well. For the hundreds of microbial species in your gut whose genomes have never been sequenced, the confetti pieces cannot be matched. The software sets them aside. Those organisms do not appear in your report.

Diagram 3.7. Who Is Represented in the Reference Data

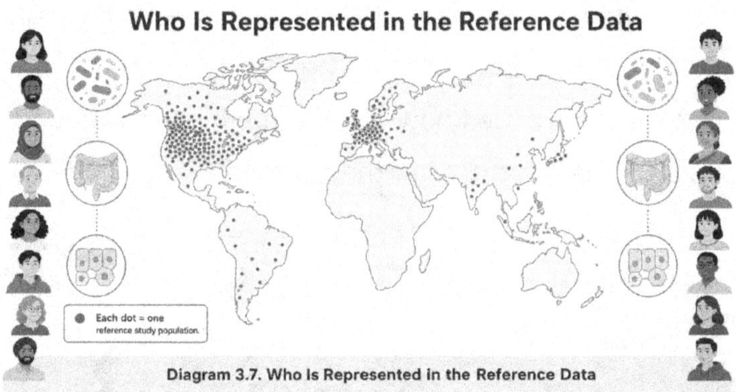

Current estimates suggest that between 50 and 80% of the microbial species in a typical human gut sample cannot be matched to any existing reference database entry. That number is improving as more genomes are sequenced and published, but even a conservative estimate is remarkable. More than half of your gut's inhabitants may be scientifically nameless, real organisms performing real biological functions that remain invisible to any sequencing-based test. The results you receive represent, at best, a partial inventory, and the visible fraction is not random. The organisms that have been cataloged are disproportionately those researchers have found most interesting, most abundant, or most closely linked to disease.

This invisibility of novel organisms is one reason microbiome scientists describe the field as a frontier. It is not a mature science with a complete map; it is more like early cartography, where the known coastlines are drawn with reasonable accuracy and vast inland territories are marked with the equivalent of "here be dragons." The result is a reading of the known world, not the whole world.

Diagram 3.8. The Library After the Shred

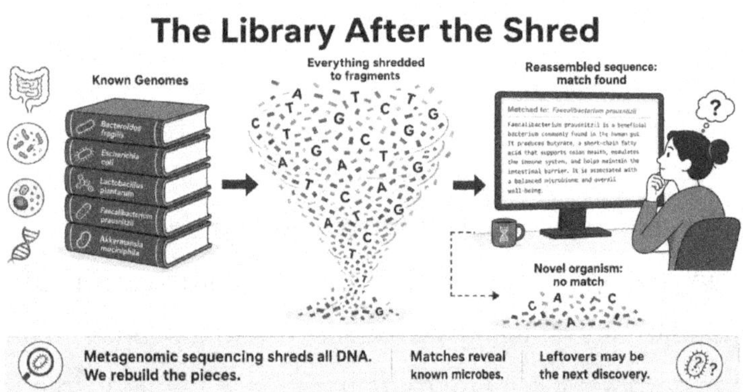

The missing-pages problem also explains why the metagenomes of two healthy people can look strikingly different from each other. If the visible fraction of the microbiome is roughly forty percent of the actual community, and the invisible remainder varies dramatically between individuals, the visible data will show substantial person-to-person differences that may have nothing to do with health outcomes. This is part of why defining a "normal" or "healthy" microbiome has proven so elusive. The field is trying to define normal from a partial dataset.

Race and Diversity in Reference Datasets

Every reference database used by bioinformatics pipelines to assign species identities, calculate diversity scores, and define what a "typical" microbiome looks like was built from samples collected by researchers working with available populations. And those populations have not been representative of humanity. The overwhelming majority of gut microbiome reference data published before 2020 came from white adults living in North America or Western Europe, most of whom were affiliated with universities as students, faculty, or staff.

The consequences are not subtle. When a consumer testing company or a research pipeline compares your sample to its reference population and tells you that your diversity score is "below average," that assessment is made relative to a reference distribution. If the reference distribution was built primarily from White American or European adults eating Western diets, then "normal" is effectively defined as resembling those populations. A person of West African descent eating a traditional high-fiber diet might receive a report suggesting their microbiome is "abnormal" in ways that actually reflect a healthier, more diverse microbial community than the reference population. The test is not measuring deviation from health; it is measuring deviation from a study population.

This matters for clinical interpretation. Early microbiome research found that people with certain conditions, including obesity, type 2 diabetes, and inflammatory bowel disease, had lower microbial diversity than healthy controls. Those findings formed the basis for a widely shared narrative that "more diversity equals better health." But if the "healthy controls" in those studies were mostly White Western university adults, diversity benchmarks derived from those studies may not apply universally. Indigenous populations, rural African populations, and communities in South Asia and Latin America consistently exhibit microbial diversity patterns that differ from those of Western urban populations in ways that reflect diet and environment rather than disease.

The scientific community has recognized this as a significant problem and has begun corrective work. Large international projects and collaborations in sub-Saharan Africa, rural South America, and Southeast Asia have expanded reference databases substantially over the last decade. But the legacy of decades of Western-skewed sampling does not disappear quickly and remains embedded in the reference databases most pipelines use today. Until those databases are genuinely globally representative, any claim about what your microbiome should look like must be understood as a claim relative to a specific, limited population.

For consumers from non-Western backgrounds, this has a direct practical implication: a result saying your microbial profile is "outside the typical range" may mean that the range was defined without people like you in the reference sample. Responsible companies are beginning to disaggregate their reference populations and report results relative to populations that share more of your biological and dietary context. This is an equity issue as well as a scientific one. The people most systematically misled by a reference-database-driven report are often those from populations underrepresented in the data, and those populations have historically had less access to high-quality medical care. Getting the reference data right is a prerequisite for making microbiome science genuinely equitable and genuinely useful.

Returning to Marcus and the Viome Report

Marcus, our composite fifty-three-year-old, has spent a Saturday afternoon going through his Viome results. The bar charts are professional and colorful, and several organisms are flagged with descriptions that sound authoritative. His "Gut Microbiome Score" is 68 out of 100, which is described as "room for improvement." His Akkermansia muciniphila is labeled "low."

Viome's metatranscriptomic approach reads which genes are being expressed, not just which organisms are present, and that functional layer adds real information relative to basic 16S sequencing. At the same time, whether Akkermansia abundance, as measured by RNA expression in a consumer stool sample, reliably predicts an individual's metabolic health remains unsettled. Akkermansia muciniphila has been associated in multiple studies with leaner body composition and better metabolic markers, but these are population-level associations based on different protocols. Whether a "low" reading in Marcus's report means the same thing as "low" in those studies is unknown.

The chia seed recommendation is also worth examining. Chia seeds appear on Marcus's "Avoid" list because his results show elevated levels of Bacteroidetes species that Viome's algorithms associate with FODMAP sensitivity in some users. Marcus has eaten chia seeds without any digestive complaints for three years. He might reasonably test the recommendation by eliminating chia seeds for a few weeks and tracking how he feels, treating it as an experiment, or he might decide that three years of comfortable consumption is sufficient evidence to be skeptical. Both responses are appropriate.

Marcus's best use of this report is as a conversation starter with a registered dietitian or physician interested in gut health, and as a prompt to notice patterns in his own digestion. The report has given him new vocabulary for thinking about his gut. It has not told him what is wrong or what will make him sick. Used with those expectations in mind, it is a reasonable tool. Used as a diagnostic document or a prescription, it overreaches what the current science can support.

Takeaways

- DNA sequencing reads the order of the four genetic letters (A, T, C, G) in a sample. Modern sequencing machines can process billions of fragments in hours, at a cost that was unimaginable two decades ago.
- Two main approaches dominate microbiome science. The 16S rRNA method targets a universal bacterial gene to quickly and cheaply answer "who is here?" Shotgun metagenomics sequences all DNA to answer both "who is here?" and "what can they do?"
- Bioinformatics pipelines transform billions of raw sequence reads into species lists, diversity scores, and functional predictions. Every step, from DNA extraction to reference database choice to statistical method, introduces variability that shapes the final result.

- Consumer test kits like Viome, Ombre, and Zoe measure real biology, but their interpretive frameworks rely on reference populations and correlational algorithms still being validated. A result labeled "low" or "abnormal" is relative to a specific comparison group, not an absolute truth.
- The same biological sample sent to two labs can produce meaningfully different results. This reflects genuine technical variability. It is a reason for appropriate skepticism about highly specific recommendations from any single test.
- Between fifty and eighty percent of microbial species in a typical gut sample cannot be matched to any existing reference database. The results any test provides are a partial inventory of the known world, not a complete map of your gut's biology.
- Most reference databases used to define "normal" gut microbiome profiles were built primarily from White Western adults. This creates systematic bias in how results are interpreted for people from other populations, and it is an active scientific and equity concern that researchers are working to address.
- Smart use of microbiome test results means treating them as hypotheses to explore and conversations to start, not as verdicts to act on. Pair results with your own experience, consult a qualified clinician for health

concerns, and maintain healthy appreciation for what the technology genuinely cannot yet tell you.

4 The Gut-Brain Highway

STRONGLY SUPPORTED: Solid epidemiological and mechanistic evidence in humans.
PROMISING: Plausible mechanism, correlational human evidence, causation open.
MARKETING-ONLY: Extrapolations from mice, single-study findings, or wellness-industry overreach.

Imagine a morning at McMaster University in Hamilton, Ontario, sometime in 2011. The laboratory is quiet except for the soft hum of filtered air and the faint rustle of mice in their enclosures. The researchers watching those enclosures are about to do something that, even by the standards of a field already full of surprises, strikes onlookers as a little outrageous. They are going to transplant gut bacteria from one group of mice into another group and watch whether personalities change. Not body weight. Not digestion. Personalities.

The mice are not random picks from the same litter. The researchers deliberately selected two strains with well-documented temperamental differences. One strain is bold and adventurous, eager to explore open spaces, quick to approach novel objects. The other is timid and cautious, preferring to huddle near walls, reluctant to venture into exposed territory. These differences are stable, reproducible, and well-known to anyone who works with these animals.

The result stops the room. The cautious mice, once colonized with gut bacteria from the bold strain, begin to act more boldly. The bold mice, once colonized with bacteria from the cautious strain, begin to behave more cautiously. The personality shift follows the microbiome. If you had told an auditorium of neuroscientists twenty years before this experiment that you could transfer personality traits by transplanting gut bacteria, the laughter would have been long and loud. By 2011, the laughter had given way to something far more interesting: a room full of researchers who could not look away.

That experiment is not the whole story of the gut-brain axis. But it is the right place to begin, because it captures exactly what makes this corner of microbiome science so fascinating and so important to understand carefully. The findings are real. The biology is real. And the temptation to leap from "mouse personality swap" to "fix your gut, fix your mind" is real too. That leap is worth resisting until we have walked through the full picture together.

The Enteric Nervous System: The Second Brain

Most people, when they think about the nervous system, picture the brain and spinal cord: a central command structure housed in the skull and the backbone. This picture is accurate as far as it goes, but it leaves out something remarkable. Coiled inside the walls of your digestive tract, from the esophagus all the way to the rectum, sits an elaborate network of approximately 500 million neurons. The human spinal cord contains roughly 100 million neurons. The enteric nervous system contains five times as many.

This network is not a passive relay station waiting for instructions from headquarters. It has its own sensory cells, its own interneurons for processing signals, and its own motor neurons for triggering muscular contractions. It regulates the wave-like movements that push food through the digestive tract, manages secretion of digestive enzymes and mucus, and monitors the chemical environment of the gut, all largely without checking in with the brain. Researchers know this because patients whose vagus nerve, the major communication cable between gut and brain, has been surgically severed can still digest food normally. The regional office keeps running even after the phone line to the capital has been cut.

The analogy of a regional government is useful here. Picture the central nervous system as a federal government: setting broad policy, managing crises, and coordinating the whole. The enteric nervous system is a capable regional administration handling local infrastructure largely on its own, connected to the capital through a dedicated communication corridor. That corridor is the vagus nerve. And the people living and working inside that regional office, influencing everything from the local water supply to the ambient mood, are the trillions of microbes inhabiting the gut.

Diagram 4.1. The Two Brains

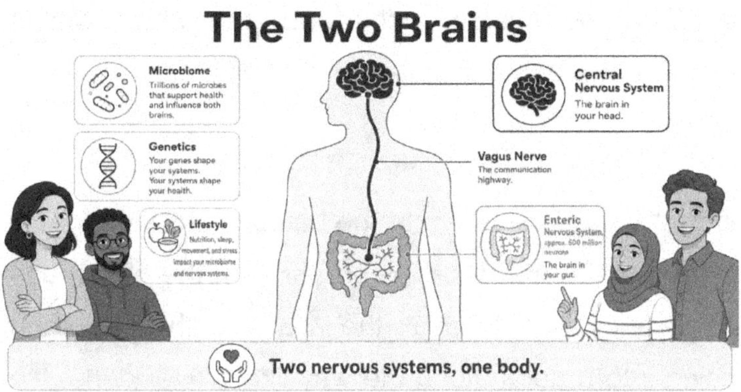

The enteric nervous system was first described in the 1800s, long before anyone understood the gut microbiome. What was not understood for another century was how profoundly gut microbes could influence the enteric network and, through it, the central nervous system itself. The gut feeling, the stomach-drop of fear, the nausea of dread: these are not metaphors. They are the felt experience of a communication system as real as any highway.

Understanding the enteric nervous system also helps explain why gut disorders and mental health conditions so often co-occur. Irritable bowel syndrome, for example, has a well-documented association with anxiety and depression. For a long time, clinicians assumed that depression caused bowel problems, or that both were independently triggered by stress. The more complete picture now includes a third possibility: that the gut and the brain are in continuous bidirectional dialogue, and that changes in one system ripple into the other with a regularity we are only beginning to map with precision.

The Vagus Nerve: The Superhighway

The word "vagus" comes from the Latin for wandering: this nerve wanders from the brainstem through the chest, past the lungs and heart, and all the way into the abdomen, branching into organs along the way like a river delta. It is the longest cranial nerve in the body. What most people would not guess is the direction of most of its traffic. Approximately 80 percent of vagus nerve fibers carry signals from the gut up to the brain, not the other way around. The gut is, in a literal anatomical sense, talking to the brain far more than the brain is talking to the gut.

Gut bacteria influence this communication in several distinct ways. Some species produce metabolites, small molecules generated as byproducts of their metabolism, that directly activate vagal nerve endings in the gut wall. Others stimulate specialized cells in the gut lining called enterochromaffin cells, which release chemical messengers in response to the gut environment. Still other bacteria influence the immune cells patrolling the gut, and those immune cells send inflammatory or anti-inflammatory signals that the vagus nerve relays toward the brain. This is a layered, multiply connected communication system with the microbiome as its source.

Diagram 4.2. The Vagus Nerve Superhighway

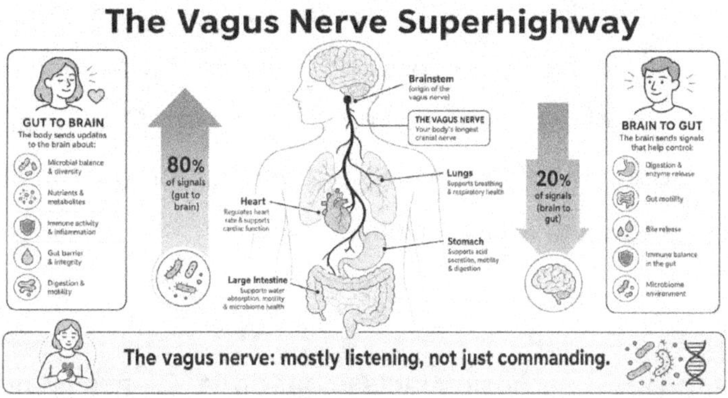

The practical importance of this architecture becomes clear when researchers experimentally block the vagus nerve. In animal studies, severing or chemically blocking the vagus nerve often eliminates the behavioral effects that would otherwise result from changes in the microbiome. When germ-free mice colonized with bacteria from anxious animals begin to show increased anxiety, that effect frequently disappears if the vagus nerve is cut first. Remove the highway and the message stops arriving at its destination. This "ablation" evidence, while impossible to replicate in healthy humans, provides strong support for the vagus nerve as a critical relay in the gut-brain conversation rather than a passive bystander.

Vagus nerve stimulation is already an established medical treatment: devices that deliver gentle electrical pulses to the nerve are used for epilepsy and treatment-resistant depression, and researchers are exploring whether similar stimulation could modulate gut inflammation in bowel diseases. The highway runs in both directions and can, in principle, be driven from either end. Whether breathing and relaxation practices that are known to activate the vagus nerve also produce downstream benefits for gut health remains an area of active investigation.

Microbes That Make Brain Chemicals

One of the most surprising facts in gut-brain science is this: approximately 90 percent of the body's serotonin is produced in the gut, not the brain. Serotonin is the chemical messenger most people associate with antidepressants and with the regulation of mood. The overwhelming majority of it is manufactured not by neurons but by the cells of the gut lining, directly influenced by the nearby microbes.

This fact requires immediate clarification to avoid a common misreading. Serotonin produced in the gut does not cross the blood-brain barrier in meaningful amounts; it cannot simply migrate upward to relieve depression. Gut serotonin primarily regulates intestinal motility and contributes to the gut's local signaling environment. But specific gut bacteria do influence serotonin production in the gut lining, and changes in those bacterial populations produce measurable changes in gut serotonin levels. The bacteria are mostly not making serotonin themselves; they are influencing the gut cells that make it, acting as conductors for a production process that uses one of the body's most important mood molecules.

Diagram 4.3. Microbes That Make Mood Chemicals

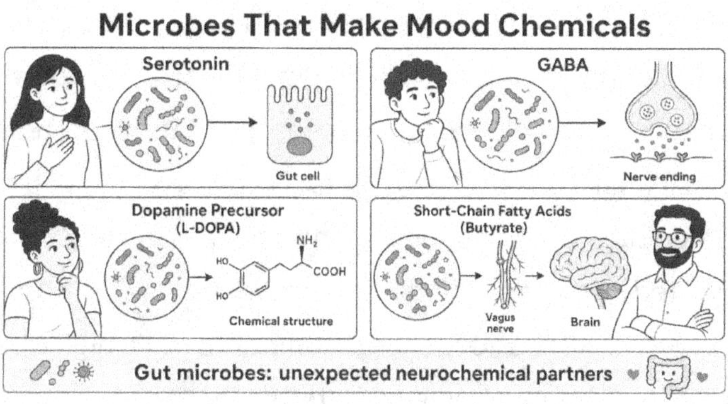

GABA, short for gamma-aminobutyric acid, tells a parallel story. GABA is the brain's main inhibitory neurotransmitter, the signal that tells excitable neurons to calm down. Low GABA activity is associated with anxiety, and many anti-anxiety medications work by amplifying GABA's effect. Certain gut bacteria, notably some Lactobacillus species studied in animal models, can produce GABA directly. Whether bacterial GABA crosses the gut wall in sufficient amounts to directly influence the central nervous system remains under investigation. What is more established is that it may activate vagal signaling locally, thereby exerting indirect effects on the brain. Dopamine's precursor, L-DOPA, is also produced by some gut bacteria, which has attracted interest from Parkinson's disease researchers, given that gut motility problems often appear years before the neurological symptoms of Parkinson's.

Short-chain fatty acids, produced when gut bacteria ferment dietary fiber, are among the best-understood molecules that gut microbes make. Butyrate, propionate, and acetate nourish the cells lining the colon, regulate inflammation, and influence the integrity of the blood-brain barrier. Animal research suggests that these molecules can modulate brain function and behavior, and that animals fed low-fiber diets show differences in brain chemistry and stress responses compared with high-fiber counterparts. The mechanisms here are better understood than for most gut-brain pathways, even if the translation to human clinical outcomes is still being worked out.

Germ-Free Mice: What They Show and Cannot Show

Germ-free mice are raised from birth in sterile, sealed chambers with no exposure to microorganisms of any kind. Their food is sterilized. The air they breathe is filtered. Their mothers carry no gut bacteria to pass on during birth or nursing. These animals live their entire lives in a state that no mammal in nature has ever experienced: a body without a microbiome.

The behavioral consequences are striking. Germ-free mice consistently show higher levels of stress-related behavior than conventionally raised mice. They are more anxious in open-field tests. They have elevated stress hormones in their blood. Their brains show measurable differences in the expression of genes related to anxiety and stress response. When researchers colonize these animals with normal gut bacteria, many of these differences normalize. The McMaster experiment described at the start of this chapter goes further: colonize a germ-free mouse with bacteria from a bold strain, and it acts bold; colonize it with bacteria from an anxious strain, and it acts anxious. The behavioral difference transfers with the bacteria.

Diagram 4.4. The Germ-Free Mouse Experiment

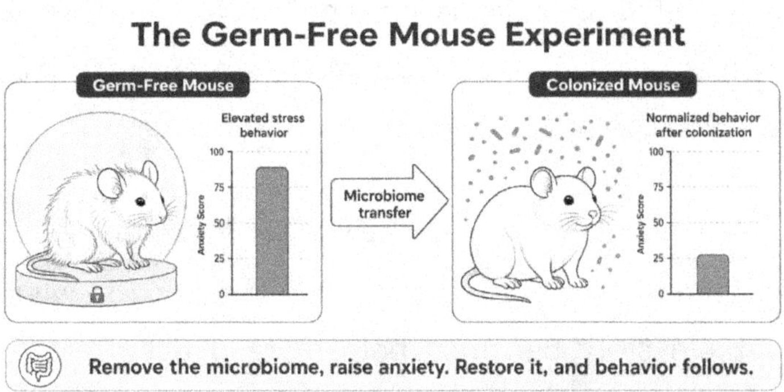

Here is where careful handling becomes essential. Germ-free mice are not simply adults who have taken too many antibiotics. They are animals that have never had a microbiome at any point in their lives. Their immune systems developed without microbial education. Their gut architecture is structurally abnormal. When researchers colonize these animals and observe their behavior normalize, they are watching a profoundly abnormal system move toward normal, not a healthy system being enhanced or degraded by a microbiome tweak.

This distinction matters for interpreting germ-free mouse data in human terms. The relevant question for your life is not "what happens when you remove a microbiome entirely?" but "what happens when a normal human microbiome shifts in composition over time?" Those are very different experiments. The germ-free mouse model demonstrates convincingly that microbes can influence brain function and behavior. It does not tell us about the magnitude of such effects in adults with established microbiomes, and the distinction matters more than most popular coverage of this research acknowledges.

Microbiome Research in Depression and Anxiety

Moving from mouse studies to human evidence is where the gut-brain story becomes both more directly relevant and more complicated to interpret. Human microbiome research on mental health has over the past decade produced findings that are intriguing, plausible, and genuinely important, alongside methodological challenges that make firm conclusions elusive.

The large-scale observational studies came first. The Flemish Gut Flora Project analyzed gut microbiomes from thousands of participants and cross-referenced those results with self-reported quality of life and depression diagnoses. The finding that attracted most attention was that two bacterial genera, Coprococcus and Dialister, were consistently depleted in people reporting depression compared to those who did not, even after controlling for antidepressant use, which itself alters the microbiome. A similar pattern emerged from the Rotterdam Study, a large Dutch population cohort, adding independent replication. Something is consistently different about the gut microbiome of people with depression, and identifying what that difference is matters for understanding the biology of the condition.

Diagram 4.5. The Evidence Pyramid

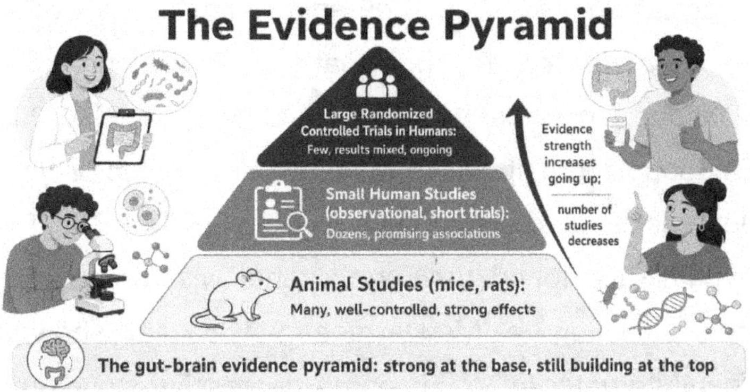

But the headline that sometimes accompanies these findings, that the microbiome "causes" depression, goes beyond what the evidence supports. What the studies show is an association: depressed people and non-depressed people have different microbiomes on average. Whether the microbial difference contributes to the depression, is a consequence of it, or reflects a third factor, such as diet or sleep, that affects both simultaneously, cannot be determined from an observational study alone. The chicken-and-egg problem is not merely rhetorical here; it is a genuine scientific challenge. Depression alters eating behavior, and eating behavior shapes the microbiome. Depression disrupts sleep, and sleep deprivation alters gut microbial populations. All of these could produce a distinctive microbiome signature in depressed populations without the microbiome playing any causal role in the depression itself.

Intervention studies using probiotics or dietary changes have produced some encouraging results. Several randomized trials found that specific probiotic combinations, particularly those containing Lactobacillus and Bifidobacterium species, reduced self-reported anxiety and depression scores compared to placebo. The SMILES trial in Australia found that participants with depression coached to follow a Mediterranean-style diet showed significantly greater improvement in depressive symptoms than a control group. These are genuinely positive results. They are also small trials with relatively short follow-up periods that have not yet been replicated at the scale required to change clinical guidelines. The evidence is strong enough to take seriously and to justify continued investment in larger trials, but not yet strong enough to support confident clinical recommendations beyond what would be advised for general health.

The Autism Connection: Promising and Contested

Few areas within gut-brain research attract as much public attention, or carry as much emotional weight, as the possibility that the gut microbiome plays a role in autism spectrum disorder. Parents of autistic children, often navigating a medical landscape that offers limited explanatory frameworks and even fewer effective interventions, are understandably drawn to any research offering new understanding. That makes precision especially important.

The starting observation is well-established: a large proportion of autistic individuals, somewhere between 30 and 90 percent depending on the study and diagnostic criteria, experience significant gastrointestinal problems. Constipation, diarrhea, bloating, and food sensitivities are reported at much higher rates in autistic populations than in the general population. This is a real clinical phenomenon that deserves far more attention and resources than it currently receives. For many families, the gastrointestinal symptoms are among the most disruptive aspects of daily life.

Research groups have reported that the gut microbiome composition in autistic children differs from that of non-autistic children, with specific genera observed at varying levels across multiple studies. Animal models have added an intriguing layer: offspring of mice with certain immune abnormalities during pregnancy exhibit autism-like social behaviors, which can be partially reversed by colonizing the offspring with specific bacterial species. A small clinical trial published in 2019 reported that a microbiota transfer therapy protocol produced improvements in gastrointestinal symptoms and autism-related behavioral measures over a two-year follow-up period, but the study was small and lacked a placebo-controlled arm for the behavioral outcomes, which makes the result promising rather than practice-changing.

These findings have generated significant hope and appropriate scientific caution in equal measure. The animal models are, even more than is typical in microbiome research, imperfect proxies for the heterogeneous human condition. The 2019 study was small, lacked a placebo-controlled arm for the behavioral measures, and the improvements could reflect natural developmental changes or relief flowing from the gastrointestinal improvements rather than any direct gut-brain effect. What is settled is that the gastrointestinal symptoms of autistic individuals are real, underdiagnosed, and undertreated, and that better gut health management is warranted regardless of any gut-brain theory. What remains uncertain is whether the microbiome plays a meaningful causal role in the neurological dimensions of autism, and families and clinicians deserve transparent updates on well-designed trials as they emerge.

What the Human Evidence Actually Shows

This section carries a different character from the rest of the chapter. It is written as an act of intellectual honesty about gut-brain research as it actually stands today, not as it is sometimes portrayed in popular media or wellness marketing. The question worth pausing on is: what does the evidence actually tell us, and where does it stop?

The gap between mouse research and human clinical evidence in the gut-brain field is substantial. Mouse studies have shown strong, often dramatic effects: colonize a germ-free mouse with bacteria from a depressed person and the mouse shows depressive-like behaviors; transfer personality-linked bacteria and the mouse's behavior follows. These effects are robust, replicable, and published in top-tier journals. They establish beyond reasonable doubt that gut microbes can influence brain function and behavior, at least in mice under controlled laboratory conditions. They do not establish that the same effects operate at the same magnitude in adult humans with complex, established, individually variable microbiomes and a lifetime of neurological development.

Human studies on the gut-brain axis are, as of this writing, largely in the associational or small-intervention phase. The observational studies showing that depressed people have different microbiomes are real and important, but they cannot establish causation. The probiotic intervention trials showing modest improvements in anxiety and mood scores are promising. Still, most have been conducted in small samples, over short durations, using varied formulations that make cross-study comparison difficult, and with outcomes measured on self-report scales subject to placebo effects. The dietary intervention trials are methodologically stronger, because diet is harder to fake than a capsule, but remain limited in number.

Diagram 4.6. Caution Box: Mouse vs Human Evidence

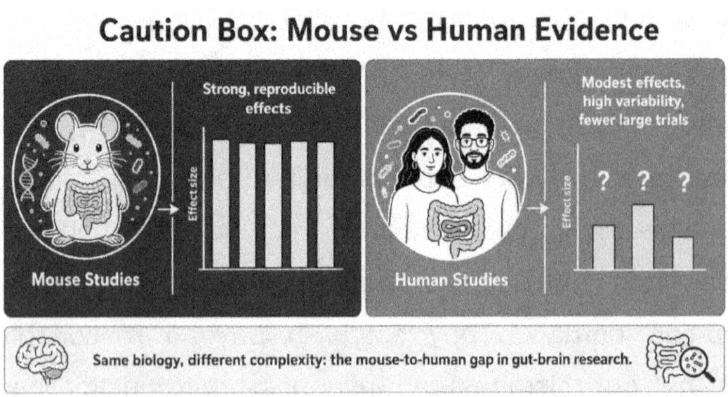

What this means in practice is that "psychobiotics," the term coined by researchers for live microorganisms with potential mental health benefits, is a scientifically legitimate area of investigation that has not yet yielded clinical-grade interventions ready for routine medical use. Certain probiotic strains have shown replicable effects in animal models and consistent, if modest, effects in some human trials. They are not yet comparable to established psychiatric treatments in terms of the quality and quantity of evidence. That may change. The field is young, the funding is growing, and larger trials are underway. The honest position today is that we do not yet know which strains benefit which people, at what doses, for how long, and for which conditions.

This calibration is not pessimism. It is the respect that good science deserves. Acknowledging the distance between the field's potential and current clinical certainty is what allows it to develop on a sound foundation, without the cycles of hype and backlash that have damaged other promising areas of medicine. The reader who understands both the excitement and the limitations is far better equipped to evaluate new findings as they arrive.

The Chicken-and-Egg Loop

Consider a composite story drawn from many real patient experiences: a woman in her mid-forties, call her Renata, who has managed mild but persistent anxiety for most of her adult life. She is not debilitated by it. She works, maintains relationships, exercises most mornings. But there is a hum of worry running beneath her days that has never fully quieted, and when life gets genuinely stressful, that hum becomes a roar. A few years ago she started reading about gut-brain research. She bought a probiotic supplement, changed her diet toward more vegetables and fermented foods, and waited.

What Renata experienced aligns with current evidence. Her digestion improved. She felt somewhat calmer after several weeks, though she was unsure how much to attribute to the probiotic, how much to the healthier diet, how much to the satisfaction of doing something active about a problem that had felt intractable, and how much to natural variation in her anxiety over time. The hum did not disappear. She kept up the dietary changes because they made her feel physically better, but she did not replace her therapy or consider the probiotic a treatment.

Renata's story illustrates one of the most important structural features of the gut-brain relationship: its bidirectionality. Stress affects the gut, and gut disturbances affect the brain, in a loop that can stabilize or destabilize depending on the direction the signals run. Chronic psychological stress activates the hypothalamic-pituitary-adrenal axis, releasing cortisol and other stress hormones. Cortisol affects gut motility, gut lining permeability, and microbial composition. Under short-term acute stress, these changes are part of the body's adaptive response. Under chronic stress lasting months or years, they can produce persistent gut dysbiosis, a less diverse microbial community, and an altered gut environment that sends amplifying signals back toward the brain through the vagus nerve and immune pathways.

Diagram 4.7. The Chicken-and-Egg Loop

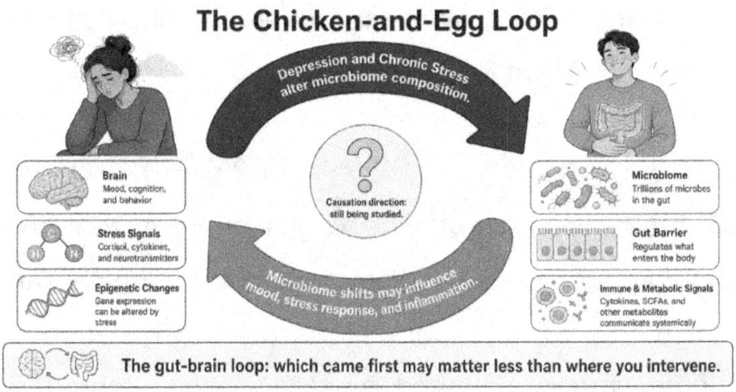

This loop is the biological basis for why stress and gut symptoms co-occur so reliably. It also explains why interventions that reduce stress often improve gut symptoms, and why improving gut health through diet sometimes reduces stress reactivity. The loop is real and bidirectional, and it is subject to intervention at multiple points. Addressing stress through therapy, diet through food choices, and sleep through consistent habits all have the potential to nudge the system in a healthier direction, even before the scientific account of each pathway is fully complete.

For many people in Renata's position, the question of which came first may matter less than the question of where the most tractable point of intervention is. For some people, addressing gut symptoms through dietary change improves mood. For others, addressing anxiety through therapy reduces gut symptoms. For many, working on both simultaneously, adding sleep and exercise, produces the most consistent improvement.

What This Means for You Today and Tomorrow

The gut-brain axis is real. The vagus nerve is real. The microbial production of neurotransmitter precursors and short-chain fatty acids is real. The association between gut microbiome composition and mental health outcomes in human populations is real. All of that is settled enough that acting on the underlying principles makes sense. What remains unsettled is precisely how to harness these mechanisms for specific clinical interventions.

The most evidence-supported things you can do today to support your gut-brain axis do not require a prescription or a specialized supplement. Eating a diverse, fiber-rich diet feeds the microbes that produce short-chain fatty acids and contribute to the microbial diversity associated with better mental health outcomes. Managing chronic stress through therapy, movement, social connection, or whatever combination works for your life reduces the cortisol-mediated disruption it reliably inflicts on the gut. Sleeping adequately, seven to nine hours for most adults, supports the restoration of microbial diversity that disrupted sleep consistently undermines. These recommendations have independent evidence of mental health, gut health, and gut-brain axis benefits.

Beyond those foundations, the emerging field of psychobiotics deserves monitoring rather than either dismissal or immediate adoption. Several research groups are conducting well-designed randomized trials of specific probiotic strains for anxiety and depression. As those results emerge over the next several years, they will begin to clarify which strains, at which doses, benefit which populations and for which conditions. When the evidence is strong enough to justify clinical recommendations, the field will say so clearly. In the meantime, probiotic supplements containing well-studied strains, manufactured to reliable quality standards, and used as complements to rather than replacements for evidence-based mental health care are unlikely to cause harm and may provide modest benefit.

Diagram 4.8. What You Can Reasonably Do Today

What You Can Reasonably Do Today

For the reader who has spent years managing anxiety, gut symptoms, or a low-level unease without understanding the biological thread connecting them, the gut-brain axis offers something valuable even before it offers a cure: a framework. It gives a biologically grounded explanation for experiences that have often been dismissed: the stomach-drop before a difficult conversation, the digestive chaos of a stressful month, the lift in mood after a week of eating well and sleeping enough. These experiences are not simply in your head. They are in your head and your gut simultaneously, because those two systems are in constant conversation through channels we are only now mapping with the precision they deserve.

The gut-brain highway is not a metaphor. It is a real road, with real signals and real consequences when the patterns go wrong. Learning to manage it through food, stress reduction, sleep, and, eventually, targeted microbiome interventions as the clinical evidence matures is part of what it will mean to take genuine care of brain health in the decades ahead. The roads themselves, the vagus nerve, the enteric nervous system, the microbial chemistry of the gut, have been there all along, working quietly whether we understood them or not.

Takeaways

- The enteric nervous system contains roughly 500 million neurons and can operate largely independently of the brain, making the gut a genuine second nervous system rather than simply a digestive organ.
- The vagus nerve carries approximately 80 percent of its traffic from gut to brain, not the other way around. The gut is, in a literal anatomical sense, sending more information to the brain than it receives back.
- Gut microbes influence brain chemistry through multiple pathways: they stimulate serotonin production in the gut lining, produce GABA and dopamine precursors directly, generate short-chain fatty acids that affect the blood-brain barrier environment, and activate immune and vagal signals that carry information toward the brain.
- Germ-free mouse studies provide compelling proof that the microbiome can influence brain function and behavior. They do not tell us directly about the magnitude of those effects in adult humans with

established, diverse microbiomes, and that distinction matters for interpreting popular coverage of this research.

- Human observational studies consistently show microbiome differences between people with depression and those without. Whether those differences are causes, consequences, or correlates of depression is not yet definitively settled. The evidence strongly supports continued investigation.
- The gut-brain relationship is bidirectional. Chronic stress disrupts the microbiome, and a disrupted microbiome can amplify the stress response. Intervening at either end of this loop, through dietary change, stress management, or sleep improvement, can shift the system in a healthier direction.
- Psychobiotics are a scientifically legitimate area of active research. They are not yet clinical-grade treatments for depression or anxiety with the evidence base of established psychiatric therapies. Use them as complements to, not replacements for, evidence-based mental health care.
- The most effective things you can do today to support your gut-brain axis- eating diverse fiber-rich foods, managing chronic stress, sleeping adequately, and working with a clinician on any significant mental health concerns- are also the behaviors most likely to produce

broad, well-documented benefits across multiple body systems simultaneously.

5 Defenders and Invaders

STRONGLY SUPPORTED: Hygiene-hypothesis association with allergy/asthma — strong epidemiological evidence across populations.
PROMISING: C-section delivery and altered microbiome trajectories — consistent signal, causation still being established.
MARKETING-ONLY: Probiotic drops as universal newborn supplements — insufficient evidence for broad clinical recommendation.

It was the autumn of 1989 when a British epidemiologist named David Strachan published a short paper in the British Medical Journal that would eventually upend the way scientists think about the immune system, cleanliness, and the invisible world of microbes. Strachan had been studying data on hay fever rates across thousands of British schoolchildren, and he noticed something surprising buried in the numbers. Children who grew up in large families with multiple older siblings had significantly lower rates of hay fever than only children or firstborns. His explanation was elegant and provocative: all those siblings brought home infections, and those infections somehow protected younger children from developing allergic disease.

The paper was only a few paragraphs long. Strachan offered a tentative hypothesis: that reduced opportunity for cross-infection in smaller, cleaner households might lead to higher rates of atopic disease, the umbrella term for allergic conditions including hay fever, eczema, and asthma. He called it the hygiene hypothesis, and the name stuck even though the underlying idea was far more subtle than "dirt is good for you."

What nobody knew in 1989, because the tools to see it did not yet exist, was that Strachan had stumbled onto the outermost edge of one of biology's most important relationships. The real story was not about infections or siblings. It was about the trillions of microorganisms that colonize the human body from the moment of birth, the ways those organisms actively teach the immune system how to function, and what happens when the conditions of modern life disrupt that education. The story that emerged is one of the most consequential in modern medicine, and it begins at birth.

Why This Chapter Matters

The immune system is not a fixed feature we are born with. It is an adaptive, learning system that develops over years in direct conversation with the microbial world around it. In the womb, the immune system is largely quiet, dependent on maternal antibodies for defense. Then birth happens, and within hours the newborn is flooded with bacteria. Those bacteria are not invaders. They are the teachers.

For most of human evolutionary history, those teachers arrived in a predictable sequence: microbes from the birth canal, then from breast milk, then from the hands and skin of people who handled the baby, then from the soil, animals, and food of the wider environment. Over the past century, and especially the past fifty years, that curriculum has changed dramatically in industrialized societies. Cesarean sections bypass the birth canal. Formula replaces breast milk. Antibiotics eliminate entire bacterial communities at critical developmental windows. Sanitized urban environments reduce contact with the ancient microbes that human immune systems co-evolved with over millennia. The consequences are written in the rising rates of allergic disease, asthma, autoimmune conditions, and inflammatory bowel disease that define modern illness in wealthy countries. This chapter is the story of that education: what it looks like when it works, what goes wrong when it is disrupted, and what the science is beginning to tell us about what we might do differently.

The First Colonizers: Birth as a Microbial Baptism

Before the first breath, before the first cry, the microbial colonization of a new human being has already begun. Passage through the birth canal is, from the microbiome's perspective, one of the most important events in a person's entire life. The vaginal canal of a pregnant woman in the final weeks of pregnancy becomes dominated by Lactobacillus species. These lactic acid-producing bacteria keep the vaginal environment acidic and inhospitable to pathogens. This Lactobacillus bloom appears to be driven by hormonal changes and seems, in the evolutionary sense, designed to ensure that the baby's first microbial encounter is with a community of bacteria that are both beneficial and protective.

As the baby travels through the birth canal, it is swabbed with maternal bacteria from head to toe. It swallows them. Within hours, those bacteria are establishing colonies in the newborn gut. The dominant early colonizers in vaginally born infants are Lactobacillus crispatus, Lactobacillus iners, and related species, alongside bacteria from the maternal gut microbiome. Within days, the initial Lactobacillus-heavy community gives way to a more complex and diverse community, but the early colonizers appear to set the tone. They prime the gut epithelium to tolerate the presence of bacteria. They begin stimulating the immune cells that line the gut wall, teaching them to distinguish between commensal bacteria, the beneficial residents that belong there, and pathogens that need to be eliminated.

The immune system of a newborn is essentially blank. It carries maternal antibodies transferred across the placenta for short-term protection, but its own adaptive immune system has not yet learned anything. Those cells need teachers, and the gut microbiome is their first and most important faculty. Within the gut wall are specialized immune structures called Peyer's patches, lymphoid follicles that sample the gut contents and present antigens to immune cells for evaluation. The process that unfolds in those structures during the first weeks and months of life is nothing less than the calibration of the entire immune system: learning which bacterial profiles are safe, developing tolerance for the commensal community, and mounting strong responses against genuine threats.

Diagram 5.1. The First Microbial Encounter

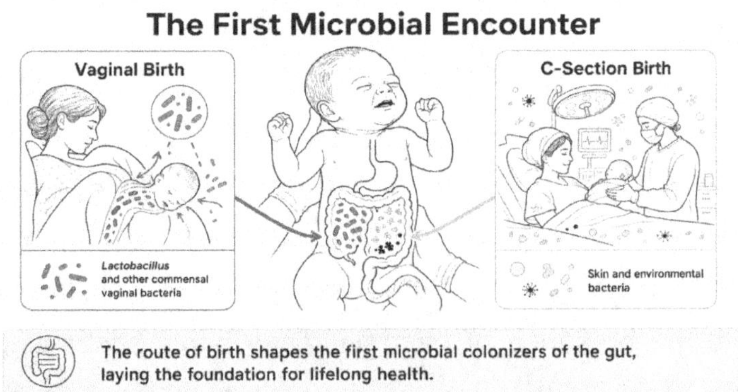

Research using metagenomic sequencing has allowed scientists to track the development of the infant gut microbiome week by week. From a relatively simple, Lactobacillus-dominated community at birth, the infant gut progresses through successional stages, eventually reaching something resembling an adult microbiome around ages 2 to 3. The introduction of solid foods is associated with a major shift in microbial community composition and a corresponding expansion of immune function. This developmental arc, from simple to complex, from reactive to calibrated, is the immune system's microbial education in real time.

What happens when that education is disrupted? The disruptions are many: the birth canal bypassed, the breast milk replaced, the early microbial community devastated by antibiotics. Each has become more common in industrialized societies over the past half century, and together they may explain much of what Strachan's small paper first hinted at: that the immune diseases of modern life are not simply genetic bad luck, but the consequence of a developmental system deprived of what it needs to function correctly.

C-Section Babies: A Different Starting Point

Consider a composite story drawn from many real patients: Elena, thirty-three years old and thirty-six weeks pregnant with her first child, learned her baby was breech and that a planned Cesarean section was the safest option. She had read about microbiome differences between C-section and vaginally born babies, and she was worried. Her obstetrician was direct: the risk to her baby from attempting a vaginal breech birth was real and immediate. The microbiome differences were real but far less certain in their long-term consequences. The C-section was scheduled for the following week.

In the United States, roughly one in three births is now delivered by C-section. Medical justifications vary: fetal distress, maternal health complications, breech presentation, prior C-section, maternal request. In many cases, C-section is genuinely the safest option for mother and baby. But from the microbiome's perspective, Cesarean delivery creates a fundamentally different starting condition for the infant's microbial education.

Diagram 5.2. C-Section vs Vaginal Birth Microbiomes

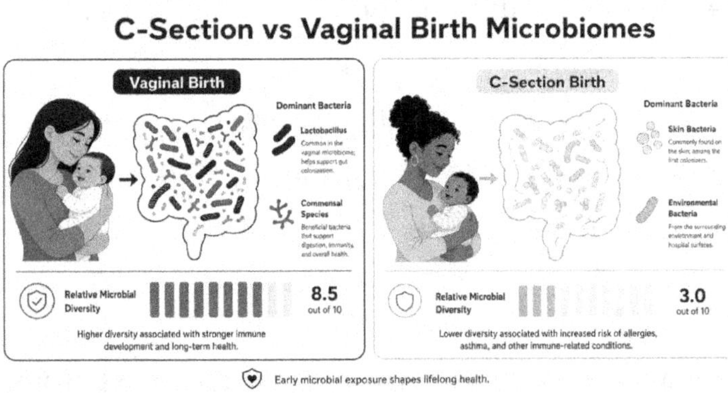

When a baby is delivered by Cesarean section, it bypasses the birth canal entirely. Its first microbial contact is not with maternal vaginal and perineal bacteria but with the skin of the medical team and the surfaces of the surgical environment. Studies comparing the gut microbiomes of C-section and vaginally born infants in the first days and weeks of life find consistent and striking differences. C-section infants are colonized first by Staphylococcus, Streptococcus, Cutibacterium, and other skin and environmental bacteria, rather than by the Lactobacillus and Bacteroides species that dominate the vaginal birth group. These are not harmful bacteria in the ordinary sense, but they are not the community that the newborn immune system appears to have evolved to learn from.

The differences in early microbiome composition are most pronounced in the first days and weeks of life, but they can persist for months and in some cases for several years. The child's microbiome typically converges toward a more typical community structure by the age of one to two years, especially in breastfed infants. Still, the window of greatest immune system training may be the very window in which the difference is largest. Large epidemiological studies, pooling data from hundreds of thousands of children across multiple countries, have found that C-section delivery is associated with modestly but consistently elevated risks of asthma, allergic rhinitis, atopic eczema, type 1 diabetes, and obesity in childhood and early adulthood. The effect sizes are not enormous, but they are robust across many populations and study designs.

This evidence does not say that C-section causes allergies or that every C-section baby will develop health problems. The population-level association is real and worth understanding, but it should not translate into guilt for parents who had medically necessary C-sections. One approach that has attracted research interest is vaginal microbiome transfer, or vaginal seeding, swabbing the newborn with gauze placed in the mother's vagina before surgery. Small studies have found this does result in a more vaginal-birth-like microbiome in C-section infants in the first weeks of life. However, evidence on clinical outcomes remains preliminary.

Breast Milk: A Living Microbial Delivery System

Breastfed infants develop differently from formula-fed infants in measurable ways: more Bifidobacterium-dominated gut communities, lower rates of allergy and asthma in many studies, and faster resolution of early infections. The question is what in breast milk drives this. The answer turns out to be more complex than simple nutrition.

Three components of breast milk do most of the relevant work. First, secretory IgA — the most abundant antibody in human milk — coats the infant's gut lining and provides passive immune protection before the infant's own immune system is fully calibrated. Second, human milk oligosaccharides (HMOs) are complex carbohydrates that the infant cannot digest; their function is to feed specific beneficial bacteria, most notably Bifidobacterium longum subsp. infantis, which has the genetic machinery to break them down. Third, breast milk itself contains hundreds of bacterial species that help seed the infant gut community. A breastfed infant receives not just food but the food those microbes need to thrive, and the microbes themselves.

Diagram 5.3. Breast Milk: Food and Inhabitants

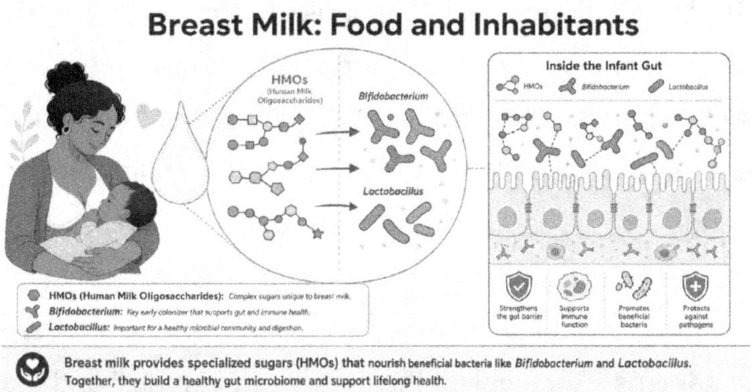

The HMO system is one of the clearest examples of co-evolution in human biology: the genome devotes substantial resources to producing molecules the infant cannot use, solely to feed specific bacterial species. Standard infant formula provides nutrition but cannot replicate this dual delivery. Formula-fed infants follow a recognizably different microbiome trajectory, with lower Bifidobacterium abundance and a less stable early community. Millions of formula-fed adults are perfectly healthy, but at the population level the differences are real.

The Hygiene Hypothesis Updated

David Strachan's original hygiene hypothesis held up well as a framing for a real observation, but ran into trouble as a mechanistic explanation. Infections in early childhood are not reliably associated with lower rates of allergic disease; some actually seem to increase allergy risk. Over the 1990s and 2000s, a more nuanced version emerged.

The most influential reframing came from Graham Rook, a British immunologist who proposed the old friends hypothesis. The cause, Rook argued, was not infections per se but the absence of microorganisms that the human immune system had co-evolved with over hundreds of thousands of years: the bacteria, parasites, and environmental microbes that have been constant companions of human beings since long before modern Homo sapiens. Rook called these the "old friends," and the hypothesis was that immune systems need regular input from these ancient companions to be properly calibrated. Without them, immune regulation fails, and the immune system becomes prone to misfiring.

Diagram 5.4. The Old Friends Hypothesis

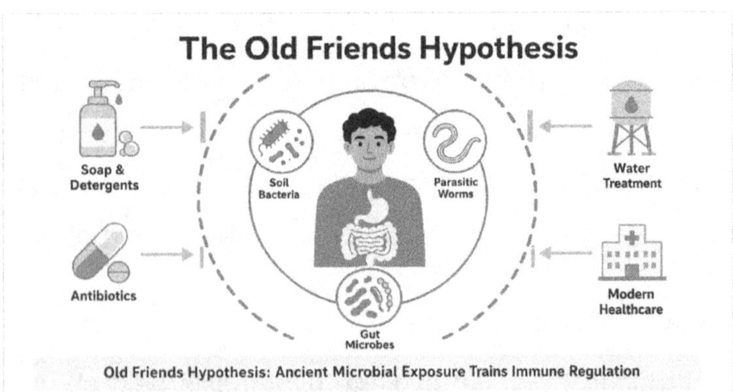

The old friends hypothesis has significant explanatory power. Hunter-gatherer populations and subsistence farming communities, groups that maintain much higher levels of microbial exposure through contact with soil, animals, and unprocessed food and water, have dramatically lower rates of allergic and autoimmune disease than urban populations in industrialized countries. This pattern holds even when genetic differences between populations are controlled for, and it holds even when comparing rural and urban populations within the same country and ethnic group.

What are the old friends? The research points to several categories: the commensal gut bacteria of Firmicutes, Bacteroidetes, Actinobacteria, and other phyla that are reduced in urban industrialized populations; environmental bacteria from soil and water like non-pathogenic mycobacteria; and parasitic worms or helminths, organisms that infected virtually all humans throughout evolutionary history and have profound immunomodulatory effects. The disappearance of helminths from industrialized countries correlates strikingly with the rise of immune diseases in those same populations.

The old friends hypothesis does not argue against sanitation. Sanitation has saved hundreds of millions of lives by eliminating waterborne pathogens and reducing infant mortality. The hypothesis argues that sanitation and the management of the built environment have, as a side effect, reduced exposure to beneficial or neutral microorganisms that the immune system needs. Spending time outdoors in natural environments, having contact with animals, eating fermented and diverse foods, and using antibiotics only when genuinely necessary may all help restore something of the old friends relationship, though the evidence for specific interventions remains an active research area rather than settled science.

Autoimmune Disease and the Microbial Deficit

Among the most striking epidemiological patterns in modern medicine is the geography of autoimmune disease. Autoimmune conditions, disorders in which the immune system mistakenly attacks the body's own tissue, include rheumatoid arthritis, multiple sclerosis, type 1 diabetes, lupus, psoriasis, and celiac disease. These conditions share a common epidemiological pattern: they are substantially more common in industrialized, high-income countries than in low- and middle-income countries, and within industrialized countries their rates have risen dramatically over the past fifty years.

The magnitude of the rise is striking. Multiple sclerosis rates in Northern Europe, North America, and Australia have approximately doubled over the past half century. Type 1 diabetes rates in Nordic children have increased fourfold since the 1950s. Inflammatory bowel disease was essentially unknown as a mass phenomenon a century ago and now affects millions worldwide. These changes represent a fundamental shift in the immune disease burden over a timescale far too short to be explained by genetic change alone.

Diagram 5.5. The Trained Immune System

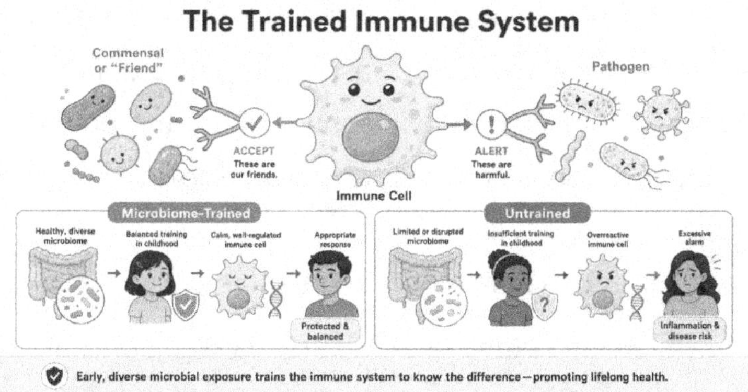

Metagenomic studies of patients with multiple sclerosis, rheumatoid arthritis, type 1 diabetes, and celiac disease consistently find lower gut microbial diversity and altered community composition compared to healthy controls. The general direction is consistent across conditions: less diversity, fewer key anti-inflammatory species, and often an overrepresentation of species associated with inflammation and barrier disruption.

A second line of evidence focuses on gut barrier integrity. The gut epithelium is a selective filter, a single cell layer with tight junction proteins that control what passes from the gut lumen into the bloodstream. When the gut microbiome is healthy and diverse, these tight junctions are well-maintained. When the microbiome is dysbiotic, the tight junctions can become leaky, allowing bacterial fragments and food antigens to pass into systemic circulation, where the immune system encounters them in contexts that can trigger inflammation and immune activation. Alessio Fasano and colleagues have conducted influential work on the role of a protein called zonulin in regulating tight junction integrity and its potential involvement in conditions such as celiac disease and type 1 diabetes.

Diagram 5.6. The Autoimmune Geography

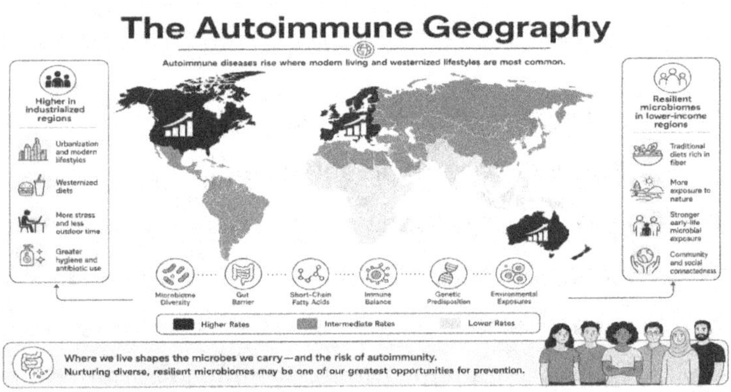

The association between microbiome dysbiosis and autoimmune disease is well established, but the causal direction is often uncertain. Autoimmune conditions may cause microbiome changes, or both may be caused by genetics or early life environment. The available data strongly suggest that a microbially rich immune education is protective and that interventions that preserve microbial diversity in early life may reduce autoimmune risk.

IBD: When the Gut Immune System Breaks Down

Inflammatory bowel disease, or IBD, represents perhaps the most direct collision between the gut immune system and the gut microbiome. Unlike allergies, which involve immune reactions to external substances, or classic autoimmune diseases, which involve immune attacks on specific self-tissues, IBD appears to involve a sustained, dysregulated immune response to the gut microbiome itself. The immune system of an IBD patient essentially fails to maintain tolerance to the commensal bacterial community that normally lives peacefully in the gut, resulting in chronic, damaging inflammation of the intestinal wall.

IBD encompasses two main conditions. Crohn's disease can affect any part of the digestive tract. However, it most commonly affects the terminal ileum and colon, involving patchy, full-thickness inflammation that can lead to strictures, fistulas, and other serious complications. Ulcerative colitis affects only the large intestine and involves continuous inflammation of the mucosal lining, typically beginning in the rectum and extending upward. Flares and remissions characterize both conditions, and both can be severely debilitating, affecting quality of life in ways that extend far beyond gastrointestinal symptoms, including fatigue, joint pain, and psychological effects.

IBD was rare in Western countries before the twentieth century and became common during industrialization. It is now seeing its most rapid growth in newly industrializing countries in Asia, South America, and the Middle East, tracking closely with urbanization and reduced microbial diversity, which strongly implicates environmental and microbial factors, not just genetics, in driving IBD risk.

Diagram 5.7. The IBD Microbiome Profile

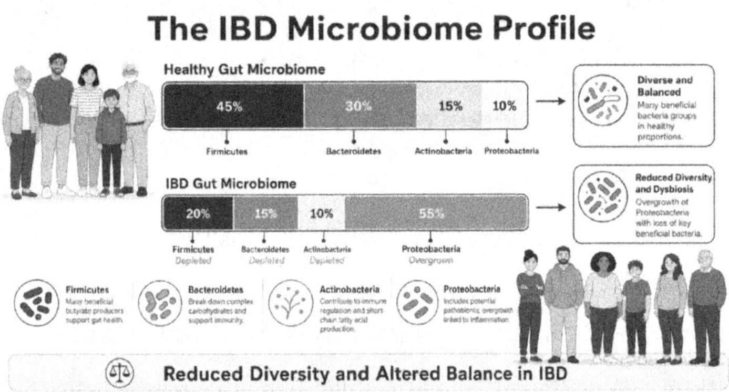

Metagenomic studies of IBD patients consistently find reduced diversity, loss of short-chain fatty acid-producing species from Lachnospiraceae and Ruminococcaceae, and expansion of pro-inflammatory species including adherent-invasive Escherichia coli. Genetic susceptibility provides the initial vulnerability, environmental triggers disrupt the microbiome, and the disrupted microbiome drives further immune activation in a self-reinforcing cycle.

Consider another composite story from many IBD patients: Andre, twenty-eight years old and recently diagnosed with Crohn's disease, asked his gastroenterologist whether changing his microbiome could cure his condition. The honest answer was that microbiome research may eventually lead to new treatments. Still, at present, no microbiome-based intervention has been shown to reliably induce or maintain remission in Crohn's disease to the extent that existing approved medications do. The right approach was proven medication alongside a varied, plant-rich diet. Andre left feeling better informed rather than falsely hopeful.

Currently, IBD research is pursuing several microbiome-based angles with genuine promise. Fecal microbiota transplantation has shown real efficacy in ulcerative colitis across multiple randomized controlled trials, with remission rates of around 30% in patients who have not responded to standard therapies. Specific engineered bacterial strains designed to produce anti-inflammatory compounds directly in the gut are in early clinical trials. Dietary interventions that systematically increase the diversity and abundance of short-chain fatty acid-producing bacteria are being studied rigorously. The field is moving from description toward intervention, testing whether restoring the microbiome can change the course of the disease.

Antibiotic Use in Childhood: The Long Shadow

Antibiotics are among the most important medicines ever developed. Before the antibiotic era, bacterial infections that are now trivially treatable- strep throat, ear infections, bacterial pneumonia, urinary tract infections- were significant causes of death and disability across all age groups. None of what follows is an argument against antibiotics. It is an argument for using them thoughtfully, because the evidence that antibiotic use in early childhood has lasting effects on the developing microbiome and the immune system it trains is now substantial and consistent.

Antibiotics do not kill only the target pathogen. Broad-spectrum drugs like amoxicillin, azithromycin, and cephalosporins kill indiscriminately across the gut microbiome, devastating communities of commensal bacteria that have nothing to do with the infection being treated. The gut microbiome of a child who has received a course of broad-spectrum antibiotics can lose a substantial fraction of its species diversity within days. While it recovers, the recovery is typically incomplete: specific species reduced or eliminated, each subsequent course adding to the accumulated disruption.

The critical point is timing. The first two to three years of life are the period of most intensive immune system training via the gut microbiome. Large epidemiological studies have found consistent associations between antibiotic use in the first two years of life and elevated rates of asthma, allergic disease, and inflammatory bowel disease in later childhood. The associations are dose-dependent: more courses of antibiotics are associated with higher risk, and broad-spectrum antibiotics appear more problematic than narrow-spectrum ones.

Diagram 5.8. Antibiotic Courses and Long-Term Risk

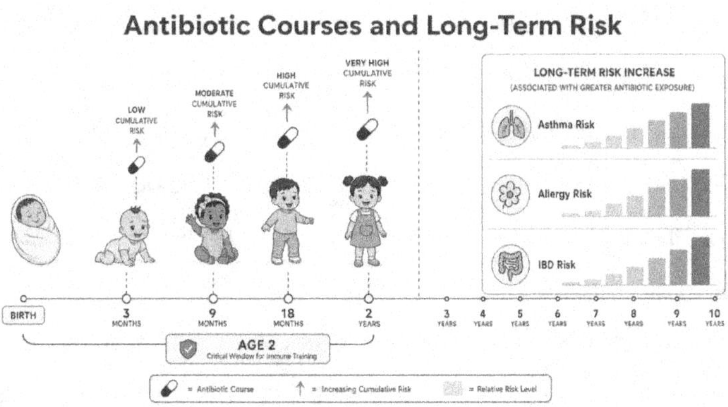

The practical implication is not antibiotic avoidance. Children with bacterial infections need treatment. The practical implications are precision: using antibiotics only when a bacterial infection is genuinely confirmed or strongly suspected, choosing narrow-spectrum agents when appropriate, and having informed conversations with pediatricians. In many countries, antibiotics are prescribed for viral conditions, particularly upper respiratory tract infections and ear infections, where they provide no benefit and carry all the risks to the microbiome. Reducing inappropriate antibiotic prescribing is a public health goal with benefits for both antibiotic resistance and microbiome health.

For parents, this means becoming comfortable asking, "Is this bacterial or viral?" Would watchful waiting be appropriate? Is this the narrowest-spectrum antibiotic that would work? Most pediatricians welcome an informed, engaged parent. After any antibiotic course, a diverse, fiber-rich diet provides the gut microbiome with the best conditions for recovery.

Elena, our expectant mother from earlier, ultimately had her Cesarean section without complication. She breastfed her daughter for eight months and asked her pediatrician to walk her through when antibiotics were and were not typically warranted. The answer was clear: ear infections in very young infants were usually treated, older children's ear infections often resolved without antibiotics, upper respiratory infections were virtually always viral. Elena left that appointment with a sense of partnership in her daughter's care. The C-section had given her daughter a different starting microbiome. Breastfeeding had significantly narrowed that gap. Thoughtful, selective use of antibiotics in the future would protect what the breastfeeding had helped build. No single factor was determinative. The whole trajectory mattered.

What This Means For You

The science in this chapter is about populations and probabilities, not certainties for any individual. Here is what it can reasonably guide for your own life and family:

- Birth method matters for the microbiome, but it is not destiny. If you or your child were born by C-section, the microbiome difference is real but not permanent. Breastfeeding, diverse diet, time in natural environments, and judicious antibiotic use all work to build and protect a healthy gut community over time.
- Breast milk is the most complete microbial support system available to a newborn. If breastfeeding is possible and desired, the evidence for its microbiome and immune benefits is strong. If it is not possible or not chosen, that is a legitimate decision, and other factors can substantially offset the microbiome consequences.
- The hygiene hypothesis, properly understood as the old friends hypothesis, argues not for living in filth but for regular contact with the natural microbial environment: time outdoors, contact with soil and animals, varied and fermented foods, and reduced reliance on antibacterial products in everyday household settings.
- Autoimmune diseases and IBD have strong environmental and microbial components alongside

genetic predisposition. Supporting a diverse, plant-rich diet that feeds a diverse gut microbiome is the most evidence-based strategy available for reducing microbiome-related immune risk over a lifetime.

- Antibiotics when needed, not when in doubt. The conversation to have with a pediatrician is not "should we avoid antibiotics" but "is this situation one where antibiotics are genuinely indicated, and is the narrowest appropriate option being used." Every unnecessary course is an avoidable disruption to the microbiome at exactly the time when the microbiome is doing important work.
- The developmental window of the first two to three years of life is the period of greatest microbiome and immune system interaction. The choices made during this window, about birth, feeding, antibiotic use, and environment, have longer echoes than choices made later. But the gut microbiome remains plastic and responsive to intervention throughout life, and it is never too late to begin supporting it.
- The rising rates of allergic, autoimmune, and inflammatory disease in industrialized countries track with specific, identifiable disruptions to the microbial environment of early childhood. That means they are, at least in principle, modifiable, and the next decade of research will make practical strategies for reducing

those disruptions considerably clearer and more actionable than they are today.

6 You Are What They Eat

Tom Spector was a university student in good health when his father, Tim Spector, a professor of genetic epidemiology at King's College London, asked him to try something unusual: eat nothing but McDonald's for ten days. Burgers, fries, chicken nuggets, soft drinks, nothing else. Before he started, Tom gave a stool sample. After ten days, he gave another. When the results came back, the number that stopped everyone in the lab was this: Tom had lost roughly 40 percent of his gut microbial diversity in less than two weeks. Not a gradual drift. A measurable collapse.

By day four, Tom had been reporting feeling worse. Sluggish, a little foggy, digestion off in ways he had not expected. The microbes that thrive on plant fiber and variety had little to work with. The populations shifted, the diversity contracted. And while ten days of burgers did not permanently ruin Tom's gut, the experiment made visible something researchers had been observing for years: the human gut microbiome responds to diet with a speed and sensitivity most of us have never been told about.

Tim Spector wrote about the experiment; it appeared in news outlets around the world and had the effect few peer-reviewed papers ever achieve. It made the connection between food and gut bacteria vivid and human. The subject was not a lab mouse. It was a healthy young man eating the food that millions of people eat every single day, watching a core part of his internal biology shrink in real time.

Why This Chapter Matters

Diet is the single most powerful modifiable factor shaping the gut microbiome over a lifetime. Genetics, birth mode, antibiotic history, and geography all leave marks, but none of them change in response to the grocery choices you make this week. Food does. Repeatedly and reliably, in ways that are now measurable.

This matters because the microbiome is not a passive bystander. The trillions of bacteria living in your large intestine are metabolically active. They break down foods you cannot digest yourself and produce compounds that travel into your bloodstream, reach your liver, and influence everything from blood sugar to immune responses to long-term risk of type 2 diabetes and fatty liver disease. What you feed those bacteria determines, in large part, what they give back to you. That exchange is the subject of this chapter.

Fiber: The Microbiome's Preferred Fuel

Virtually every research team studying the gut microbiome, regardless of which disease they focus on or which population they work with, arrives at the same basic recommendation: eat more fiber. Not the same fiber, not supplements in particular. Just more indigestible plant material that the human digestive system cannot break down on its own. The convergence is not about nutrition ideology. It is about what gut bacteria are built to do.

When you eat whole grain bread, lentils, or oats, you absorb what you can in the small intestine: simple sugars, some proteins, fats, vitamins. But a substantial portion of the plant material, particularly the complex carbohydrates bound up in plant cell walls, passes through the small intestine largely intact. These are the dietary fibers: pectin from fruits, cellulose from vegetables, inulin from onions and garlic, resistant starch from beans and unripe bananas. By the time this material arrives in the large intestine, your own digestive enzymes have done nothing with it, essentially.

That is where the microbiome takes over. The bacteria of the large intestine produce carbohydrate-active enzymes, called CAZymes, that can dismantle fiber structures the human body cannot touch. Different bacterial species have different specializations. What they share is the ability to extract energy from the indigestible and to produce something valuable in return: short-chain fatty acids, or SCFAs.

SCFAs are small organic acids produced as byproducts of bacterial fermentation. The three most important are butyrate, propionate, and acetate. Butyrate is the primary energy source for colonocytes, the cells lining the colon wall. Without adequate butyrate, colonocytes weaken, the gut barrier becomes more permeable, and the immune system encounters a steady trickle of microbial products that should have stayed inside the gut. Butyrate also has potent anti-inflammatory properties. Propionate travels to the liver, where it participates in regulating glucose production and cholesterol synthesis. Acetate circulates in the bloodstream as fuel for muscles and other tissues and acts as a signaling molecule that influences appetite and fat storage.

Diagram 6.1. Fiber Becomes Fuel

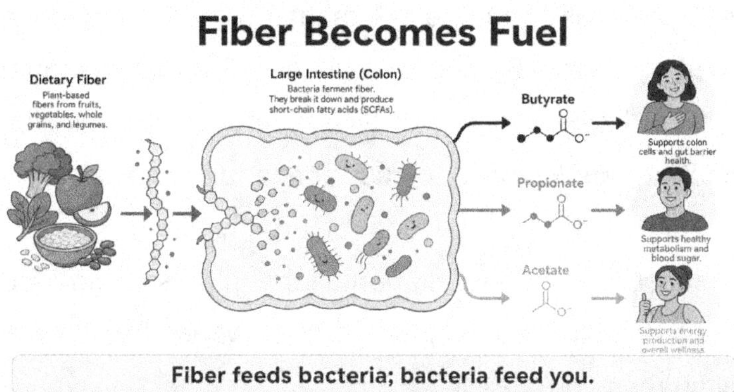

People in industrialized Western countries consume roughly 10 to 15 grams of dietary fiber per day on average. Traditional populations in rural Africa and parts of Asia and South America typically consume 40 to 60 grams or more. Most dietary guidelines recommend 25 to 38 grams daily, a target that the vast majority of Western populations never reach. The consequence is not subtle. Low-fiber diets starve the microbiome. Bacteria that specialize in fiber fermentation decline, SCFA production drops, the gut barrier weakens, and inflammation rises. Justin Sonnenburg at Stanford has described this with the image of a garden: stop watering it and the plants die back. Fiber is the water. A Western low-fiber diet is, from the gut microbiome's perspective, a prolonged drought.

The type of fiber and the quantity matter as well. Inulin and fructooligosaccharides, found in garlic, onions, leeks, and asparagus, selectively feed Bifidobacterium and Lactobacillus. Resistant starch, found in cooked-and-cooled potatoes, unripe bananas, and legumes, preferentially feeds Ruminococcus and Faecalibacterium prausnitzii, both prolific butyrate producers. Pectin, abundant in apples and citrus, feeds different species again. This is one of the central arguments for dietary variety: it is not just about nutrients for you. It is about food for thousands of microbial species whose combined activity shapes your health.

Fermented foods sit in a related but distinct category. Yogurt, kefir, sauerkraut, kimchi, and miso contain live bacteria and organic acids that interact with the resident microbiome in ways still being worked out. A 2021 randomized controlled trial from the Sonnenburg lab found that a diet high in fermented foods increased microbiome diversity and reduced markers of inflammation more than a high-fiber diet alone over the short term. These findings need to be replicated in larger, longer-term studies. But the signal added fermented foods to the list of dietary elements that appear to support a resilient microbiome.

The 30 Plants a Week Research

When researchers started asking which dietary factor most strongly predicted microbiome diversity, the answer was not total caloric intake, protein quantity, fat type, or even total grams of fiber. The factor that kept appearing was variety: the number of different plant foods a person ate each week.

The data came from the American Gut Project, a citizen science initiative that analyzed stool microbiome samples from over 11,000 participants in the United States, United Kingdom, and Australia. Participants submitted detailed dietary questionnaires alongside their samples. When researchers sorted participants by the number of different plant foods they ate per week, a consistent pattern emerged. People who ate more than 30 different plant foods per week had significantly higher gut microbial diversity than those who ate fewer than 10, and this held regardless of whether participants were vegan, vegetarian, or omnivorous.

Diagram 6.2. The 30 Plants a Week Goal

Why does variety matter so much? Different bacteria are adapted to process different plant compounds. A microbiome fed on a narrow range of plants can support only the species able to survive on those substrates. A microbiome fed on 30 or more different plants has far more ecological niches to fill. Bacteria that specialize in brassica fermentation coexist with bacteria that process legume fibers and those that handle berry polyphenols. The ecosystem becomes richer and more resilient.

What counts as a "plant" is worth clarifying, because 30 can seem intimidating until you see the full scope. Fruits count individually: an apple is one, a banana is two, blueberries are three. Vegetables, legumes, and whole grains each count individually. Nuts and seeds each count. Herbs and spices count: cumin, turmeric, black pepper, cinnamon, oregano each contributes a unique set of plant compounds; tea, coffee, and dark chocolate count. A week of thoughtful eating often yields 25 or more plants without major effort; adding five more is straightforward.

The 30 plants message has had unusually wide reach because it is specific, positive, and additive. It does not tell you to eliminate anything. It does not require expensive supplements. It asks you to add. That framing resonates in a way that restrictive dietary advice typically does not. It is also worth noting that frozen, canned, and dried plant foods all count, which matters for budget and access. The framework is genuinely available across income levels with some planning.

Polyphenols: Plant Colors That Feed Bacteria

The deep purple of blueberries, the rich red of pomegranate arils, the dark brown of cacao nibs, the green-gold of extra virgin olive oil: these colors arise from polyphenols, a vast family of plant compounds with an unusual property. A significant fraction passes through the small intestine largely unabsorbed, arriving in the large intestine intact, where the gut microbiome metabolizes them into bioactive compounds their original forms could not produce.

Polyphenols are the pigments, the bitter notes, and the astringent qualities in plant foods. Flavonoids in berries, citrus, tea, and onions. Stilbenes like resveratrol in grapes. Lignans in flaxseeds. Tannins in dark chocolate and pomegranates. Phenolic acids in coffee, one of the richest sources of polyphenols in many Western diets. These compounds serve as defenses for the plant against pathogens, UV radiation, and insects. For us, they offer something different.

Diagram 6.3. Polyphenols in Plant Colors

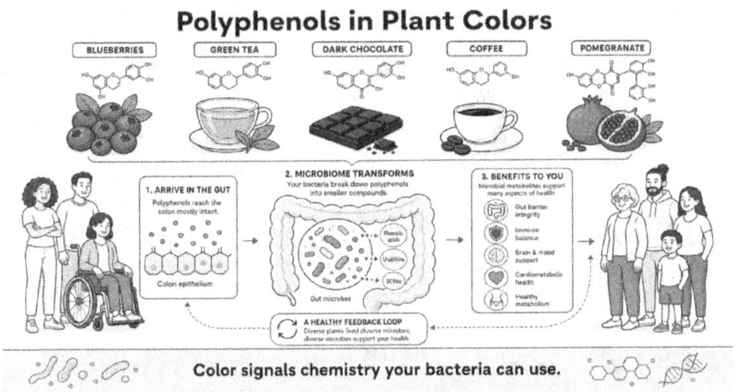

Gut bacteria that process polyphenols convert them into metabolites the body can absorb and use in ways the unmetabolized compounds never could. Equol, produced from soy isoflavones by certain gut bacteria, has estrogen-like activity. Urolithins, produced from ellagitannins in pomegranates and walnuts, have been studied for effects on mitochondrial function and inflammation. Phenolic acids derived from coffee polyphenols have anti-inflammatory and potentially neuroprotective properties. The conversion of polyphenols into these active forms depends on which bacteria are present, which is why the same polyphenol-rich food can produce different effects in different people.

This is also why people who eat polyphenol-rich plants but have low microbiome diversity may not be getting the full benefit. The bacteria doing the conversion have to be there in sufficient numbers. Feeding a diverse, healthy microbiome polyphenol-rich foods is a multiplier; giving polyphenol-rich foods to a depleted, low-diversity microbiome is like sending ingredients to a kitchen with very few cooks.

The epidemiological evidence for polyphenol-rich diets is compelling. The Mediterranean dietary pattern, among the most studied and best-supported dietary frameworks for cardiovascular and metabolic health, is extraordinarily rich in polyphenols: olive oil, tomatoes, eggplant, herbs, legumes, dark berries, coffee, and dark leafy greens. Similar patterns of high polyphenol intake appear in traditional diets from Okinawa, parts of Sardinia, and regions of Central America with unusually low cardiovascular disease rates. The common thread is not a single superfood. It is persistent, high-variety exposure to plant pigments across a lifetime. Polyphenol supplements have not replicated the strength of this food-based evidence; whole foods deliver polyphenols alongside other plant compounds that support more complete bacterial metabolism.

Personalized Glycemic Response: You Are Not Average

Consider a composite story drawn from a well-documented research finding. Two women, both in their early fifties, both in generally good health, each eat the same standardized meals for two weeks while wearing continuous glucose monitors. One meal includes white rice. After eating it, the first woman's blood sugar rises modestly, peaks within an hour, and returns to baseline. The second woman's blood sugar spikes sharply, stays elevated much longer, and falls slowly. By every standard glycemic index measure, she just experienced the metabolic equivalent of a candy bar. Same rice. Same portion. Completely different outcome.

This is not a fictional experiment. It reflects what Eran Segal and Eran Elinav at the Weizmann Institute of Science found when they attached continuous glucose monitors to 800 Israeli adults over two weeks and measured blood sugar responses to hundreds of standardized meals. The variability in responses to identical foods was enormous, far larger than the glycemic index would predict. When the researchers examined explanatory factors, the gut microbiome was among the strongest predictors, alongside fasting glucose levels, body mass index, and meal timing.

Diagram 6.4. Same Food, Different Responses

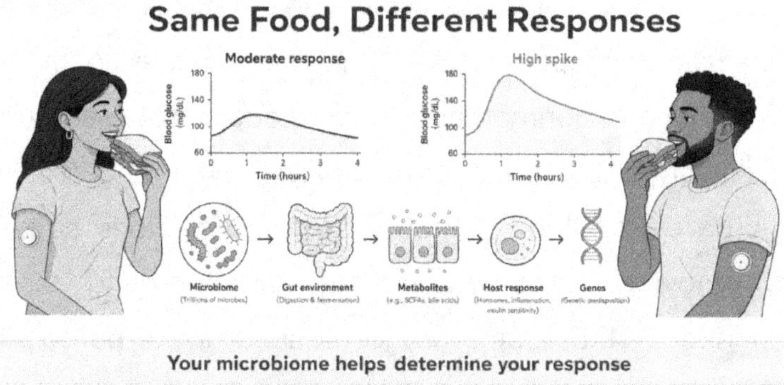

The mechanism involves multiple pathways. Some gut bacteria produce SCFAs that improve insulin sensitivity and moderate glucose uptake from the gut. Others produce metabolites that impair insulin signaling. Some bacterial populations slow gastric emptying; others speed it up. The ratio of different bacterial genera, the enzymes they express, and the metabolites they produce collectively shape what happens to your blood sugar after every meal in ways no glycemic index table can capture, because that table was constructed from average population responses. You are not an average.

Segal and Elinav developed a prediction algorithm using microbiome composition and other biological data to forecast individual glycemic responses more accurately than the standard glycemic index. In a clinical trial, participants who received personalized dietary recommendations based on their microbiome profiles achieved better blood sugar control than those who received standard low-glycemic-index advice. Routine personalized microbiome-based nutrition advice remains expensive and experimental for most people, but research is definitively shifting the conversation. When someone tells you a certain food is "good" or "bad" for blood sugar without knowing anything about the person asking, they are offering a statistical average dressed up as personal advice.

In practice, paying attention to your own responses to food matters. Wearable continuous glucose monitors have become available to consumers in many countries, and early users consistently report surprises: foods they assumed were harmless cause spikes, and foods they had been told to avoid sit well. The research also reinforces the value of microbiome diversity, because people in the Weizmann study with more diverse microbiomes showed more moderate and more predictable glycemic responses on average.

The Western Diet: An Ecological Disruption

Picture the ecosystem of a tropical rainforest: thousands of species occupying thousands of niches, each contributing to the stability of the whole. Now picture most of the forest cleared and replaced with a monoculture crop. A few highly adapted species thrive. Most others disappear. The ecosystem becomes less diverse, less redundant, and more vulnerable. A single blight can devastate a monoculture in a way that would never cause a functioning rainforest to collapse.

The gut microbiome of someone raised on a traditional, plant-rich diet from communities like the Hadza hunter-gatherers of Tanzania, rural populations in West Africa, or the highlands of Papua New Guinea, looks something like the rainforest. Researchers collecting stool samples from these populations consistently find gut microbiome diversity far exceeding that typical in Western populations: more species, more even distribution, stronger representation of fiber-fermenting bacteria, higher SCFA production, and lower inflammatory markers. The gut microbiome of someone raised on a typical Western diet is often more like a monoculture: a narrower range of species, dominated by a few that thrive under conditions of low fiber, high fat, and high levels of refined sugar.

Diagram 6.5. Western Diet vs Traditional Diet Diversity

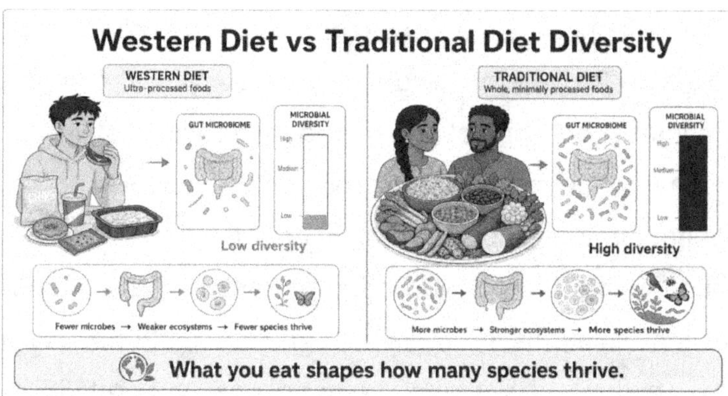

The Western diet has a recognizable signature: high in processed foods, refined grains, added sugars, and industrially produced fats; low in dietary fiber, polyphenols, and fermented foods. In feeding study after feeding study, this pattern produces consistent microbiome changes: reduced diversity, reduced abundance of butyrate-producing bacteria, and increased abundance of bacteria associated with inflammation and intestinal permeability.

Some of the most revealing data comes from studies tracking immigrants as they transition from traditional diets to Western ones. Research following Southeast Asian immigrants to the United States found that within months of arrival, gut microbiome composition had shifted measurably toward Western patterns. The longer participants had lived in the United States, the more pronounced the shift. Studies of the second generation showed microbiomes that still looked more Western. These findings correspond to well-documented increased rates of obesity, type 2 diabetes, and inflammatory bowel disease seen in immigrant populations after a generation or two.

The comparison with traditional diets requires care. Traditional diets are not universally optimal. Food security, parasitic infection rates, and other variables shape health in these populations in ways that dietary patterns alone cannot capture. What the comparisons support is a narrower claim: the long-term reduction in dietary fiber, variety, and fermented foods that characterizes industrialization has measurable consequences for the gut microbiome, aligning with rising rates of conditions associated with chronic inflammation. The good news is that the gut microbiome is not permanently locked into any pattern. Dietary changes produce measurable microbiome shifts within days to weeks, and sustained increases in plant variety yield positive changes regardless of where someone starts.

Obesity, Diabetes, and the Microbiome

The mouse experiments that first made the gut microbiome's connection to obesity undeniable were striking in a way that even non-scientists could grasp. Take a germ-free mouse, raised in sterile conditions with no gut microbiome, and transplant gut bacteria from an obese donor mouse into it. The recipient gains significantly more weight than a germ-free mouse given gut bacteria from a lean donor, even when both mice eat the same amount of the same food. The microbiome transplant alone is sufficient to reproduce a core feature of obesity: excess fat gain on a given caloric intake.

The leading mechanism involves how different microbial communities extract energy from food. Certain bacterial communities appear to extract more usable calories, leaving less in the stool. They also produce different SCFA profiles and different effects on the hormones that regulate hunger and satiety. Ghrelin, leptin, GLP-1, and PYY, the hormones that govern when you feel hungry and when you feel full, are all influenced directly or indirectly by microbial metabolites. A gut microbiome that chronically disrupts satiety signaling makes overeating easier and weight regulation harder, independent of willpower.

Diagram 6.6. The Obesity Microbiome Transplant Mouse

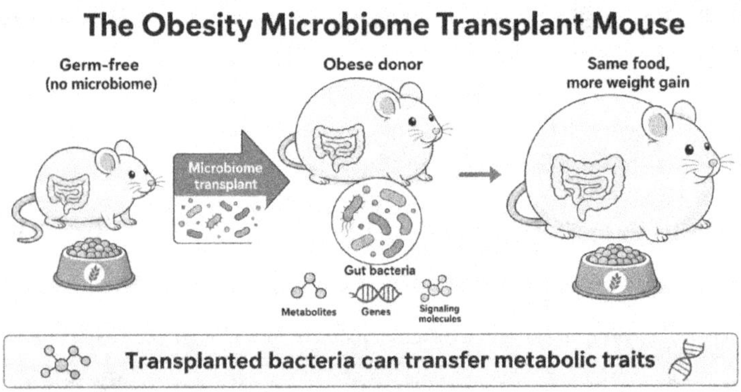

The human evidence is more complicated. The Firmicutes-to-Bacteroidetes ratio, which was robust in early mouse studies, has proven inconsistent in human research. Some studies find it, others do not, and the ratio is probably too simple a summary of a vastly complex ecological signal. What the human research does support, with increasing confidence, is that the microbiome is not a silent observer of obesity. It is a participant. Whether it is a primary driver, a significant amplifier, or a consequence that then feeds back into the process is a question different studies answer differently. The honest answer is that it is probably some degree of all three, varying from individual to individual.

For type 2 diabetes, the microbiome story is somewhat clearer at the mechanistic level. Gram-negative bacteria in the gut have cell walls containing lipopolysaccharide (LPS), a molecule that, when it leaks into the bloodstream through a permeable gut barrier, triggers systemic low-grade inflammation that impairs insulin signaling. Reduced butyrate production by a fiber-deprived microbiome further weakens the gut barrier, increasing LPS leakage. Propionate from bacterial fermentation suppresses hepatic glucose production; when propionate is low, that check weakens. Each of these mechanisms is a thread in a larger fabric of microbiome-metabolic interaction that makes the development and progression of type 2 diabetes more or less likely.

Fecal microbiota transplantation (FMT) research for metabolic disease is still in its early stages. A small number of human trials have transplanted gut bacteria from lean, metabolically healthy donors into people with insulin resistance or metabolic syndrome. Results have been mixed, with some studies showing short-term improvements in insulin sensitivity that faded over time, while others showed little effect. FMT for obesity and diabetes is a research direction, not a clinical reality. What is clear enough to act on is that dietary changes increasing fiber intake, reducing ultra-processed food, and adding variety are associated with microbiome changes that move in the direction of better metabolic health. Not a cure, but a meaningful and accessible lever.

The Gut-Liver Axis

Most people have a reasonably clear mental model of the liver. It filters blood, processes alcohol, and makes bile for fat digestion. What most people do not have is a clear model of how the liver continuously and intimately communicates with the gut microbiome, through a pathway so direct that it has acquired its own name in the scientific literature: the gut-liver axis.

The anatomical basis is the portal vein, a large blood vessel that collects blood draining from the entire digestive tract and delivers it directly to the liver before it enters general circulation. Everything the gut produces, absorbs, or leaks into the bloodstream passes through the liver first. Every SCFA produced by bacterial fermentation, every metabolite generated from polyphenols, every fragment of LPS from a leaky gut: the liver sees all of it before the rest of the body does.

Diagram 6.7. The Gut-Liver Highway

The Gut-Liver Highway

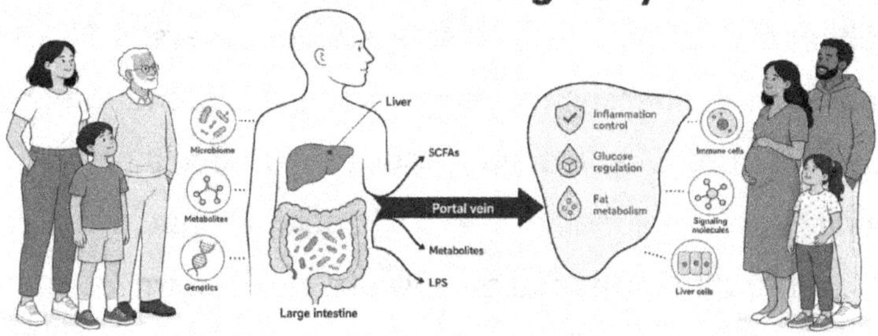

The gut talks to the liver first

The liver responds to these signals in ways that directly shape metabolic health. Propionate suppresses de novo lipogenesis and reduces the liver's output of glucose into the bloodstream. SCFAs bind to receptors on liver cells that regulate inflammation and immune activation. When SCFA production is robust, these liver processes are well-calibrated. When it falls, the calibration suffers. LPS leakage activates Toll-like receptor 4 (TLR4) on liver cells, triggering an inflammatory cascade that, when it occurs chronically at low levels, contributes to hepatic insulin resistance and fat accumulation. This process is one of the proposed mechanisms underlying non-alcoholic fatty liver disease (NAFLD), which affects an estimated 25 percent of the global adult population.

NAFLD exists on a spectrum from simple steatosis (fat accumulation without significant inflammation) to non-alcoholic steatohepatitis (NASH), which involves inflammation and cell damage and can progress to cirrhosis and liver failure. Studies comparing the gut microbiome of people with simple steatosis versus NASH find consistent differences: people with NASH tend to have lower microbial diversity, lower Faecalibacterium prausnitzii, and higher levels of bacteria that produce ethanol through fermentation, which the liver must then process as it does alcohol.

The practical implication of the gut-liver axis is that dietary changes that benefit the microbiome also directly benefit the liver. Increasing dietary fiber increases SCFA delivery through the portal vein. Reducing ultra-processed food reduces the inflammatory load a dysbiotic, permeable gut sends to hepatic cells. Adding variety and plant-rich foods supports the bacterial populations that generate protective metabolites. The liver is not a separate consideration from the microbiome. In a very direct anatomical sense, it is its downstream neighbor.

A Case Story: Finding Ground in the Noise

Consider a composite story drawn from the experiences of many patients navigating metabolic health and diet advice: Greg is 50 years old, a high school history teacher, mildly overweight, and recently told by his doctor that his fasting glucose is in the pre-diabetic range. He exercises moderately, coaches the school's cross-country team, and has been trying to eat better for years. The problem is that "eating better" has meant something different every time he looks it up. Low-fat. Low-carb. Keto. Intermittent fasting. Paleo. Mediterranean. Plant-based. His brother swears by a carnivore diet and has lost 20 pounds. His wife's colleague lost 30 pounds on a high-protein plan. And now his doctor has added a new layer of anxiety by mentioning that his gut bacteria might be part of the picture.

Greg represents a vast number of people who arrive at this juncture, not for want of motivation but for want of clarity. The science of the gut microbiome does not instantly simplify his situation. But it offers a framework that cuts through some of the noise. The question is not which dietary approach is correct. It is what his gut microbiome needs to function well, and the answer is less contested than the diet wars suggest.

What Greg's microbiome needs, as supported by the convergence of evidence in this chapter, is more fiber from a wider variety of plant sources, fewer ultra-processed foods, and probably some fermented foods added to his regular diet. His prediabetic blood sugar levels do not mean that his microbiome's influence on his metabolic health is irreversible. The gut microbiome's responsiveness to diet means that the decisions he makes at the grocery store this week will be reflected in his gut bacterial populations within days to weeks. Those populations affect the chemistry his liver receives through the portal vein, the satiety signals he gets after meals, and the degree of low-grade inflammation in his bloodstream.

The most useful thing anyone can tell Greg is this: the evidence does not point to any single magic food or dietary pattern. It points toward a set of principles that appear consistently across research traditions and populations. Eat a wide variety of plant foods. Prioritize fiber from whole food sources. Include colorful, polyphenol-rich produce. Add a fermented food or two. Reduce ultra-processed foods that crowd out the variety the microbiome needs. These principles are compatible with low-carbohydrate, Mediterranean, and flexitarian approaches alike. They sit below the level of any specific dietary label.

Greg does not need to track 30 plants per week in a spreadsheet, at least not at first. He needs to understand why variety matters so he can make better choices. He needs to know that his blood sugar response to food is not the same as anyone else's, so he can stop comparing himself to his brother and start paying attention to his own body. And he needs to know that what he does consistently over months matters far more than any single day, because the microbiome is an ecosystem that responds to trends, not individual meals.

Diagram 6.8. A Practical Plate

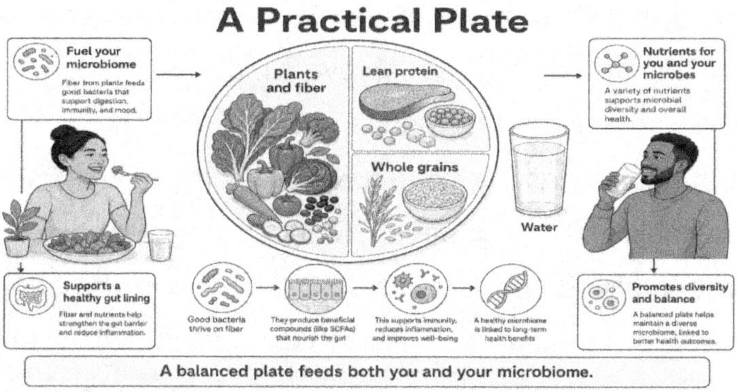

The story of Greg does not end with a dramatic transformation. It ends with a more confident understanding of why the choices he makes matter, backed by science specific enough to be useful. That is what this chapter offers: not a diet plan, but a coherent biological story that turns food choices from a source of anxiety into a source of genuine agency.

Takeaways

- The gut microbiome ferments dietary fiber into short-chain fatty acids (butyrate, propionate, acetate) that fuel the gut lining, regulate blood sugar, reduce inflammation, and communicate with the liver and immune system. Fiber is not just a digestive aid; it is the primary fuel for a biological system that influences health across the body.
- Variety in plant foods drives microbiome diversity. Large citizen science data found that eating more than 30 different plant foods per week is one of the

strongest predictors of a diverse, resilient gut microbiome, regardless of overall dietary pattern or label. Plant diversity supports bacterial diversity.

- Polyphenols, the pigment compounds in colorful fruits, vegetables, teas, coffee, and dark chocolate, reach the large intestine where gut bacteria transform them into bioactive metabolites that reduce inflammation and support cardiovascular and metabolic health. Choosing colorful foods is one of the most accessible ways to support this system.
- Two people eating the same meal can experience dramatically different blood sugar responses, and the gut microbiome is one of the strongest predictors of individual variation. Standard glycemic index advice is based on population averages, not individual biology. Paying attention to your own responses to food is more valuable than following universal food rankings.
- The Western diet, high in processed foods and low in fiber and variety, consistently reduces gut microbial diversity, increases populations of pro-inflammatory bacteria, and disrupts the production of microbial metabolites that support metabolic health. Shifts appear within days to weeks of changing eating patterns.
- The microbiome is linked to obesity and type 2 diabetes through multiple pathways: energy extraction

efficiency, appetite hormone regulation, gut barrier integrity, and the delivery of inflammatory signals to the liver. These connections add a biological layer to metabolic health management that standard advice has largely overlooked.

- The gut-liver axis describes the direct portal vein connection through which every metabolite produced by gut bacteria reaches the liver before anywhere else in the body. SCFA production supports healthy liver function; gut barrier disruption and LPS leakage contribute to hepatic inflammation and fat accumulation. Supporting the microbiome is, in a direct anatomical sense, supporting the liver.

7 The Antibiotic Paradox

Consider a composite story drawn from many real patients: Margaret, a seventy-three-year-old retired teacher from Baltimore, went into the hospital for a hip replacement surgery in the spring of 2019 in good general health, still walking two miles most mornings. The orthopedic team described the procedure as routine. It was.

What nobody told Margaret was that the broad-spectrum antibiotics she received before and after surgery, standard prophylaxis against surgical site infection, would set off a chain of events consuming the next four months of her life. By the time she left the hospital, a bacterium called Clostridioides difficile, which had been living harmlessly in her gut for years, had seized the ecological opportunity created by the antibiotic assault on her intestinal microbiome. It was multiplying at a rate her remaining defenses could not contain, releasing toxins that inflamed and stripped the lining of her colon. She was readmitted eleven days after discharge.

Over the following months, Margaret received four more antibiotic courses, each one temporarily suppressing the C. diff infection but simultaneously deepening the damage to the gut microbial community that was the only long-term barrier between her and the next relapse. Her gastroenterologist explained the paradox plainly: the tools they were using to fight the infection were, in a mechanical sense, making her more vulnerable to the same infection returning. Margaret eventually recovered after a novel treatment approach was introduced, and she agreed to let her experience be woven into composite accounts like this one.

Why This Story Matters

Margaret's experience is not a medical outlier. Clostridioides difficile infections affect nearly half a million Americans every year, and the majority of cases are directly linked to antibiotic use in healthcare settings. But C. diff represents only the most visible tip of a much larger phenomenon. For every patient who develops a clinically recognized infection after antibiotics, millions more experience a quieter disruption: a gut microbial community that has been thinned, destabilized, and in some cases permanently altered in ways that may not produce recognizable symptoms for months or years.

This chapter takes on one of modern medicine's most genuinely difficult paradoxes. Antibiotics are among the most transformative and life-saving inventions in human history. Before penicillin, a child with strep throat could die of rheumatic heart disease. The global burden of suffering that antibiotics have prevented is nearly impossible to calculate. At the same time, decades of overprescribing have produced consequences that metagenomics is now allowing us to see with painful clarity. The collateral damage to the microbiome, once invisible because we lacked the tools to measure it, turns out to be extensive, ecologically complex, and in some cases lasting.

How Antibiotics Work and Why They Are Not Precision Tools

Antibiotics are chemical compounds that kill bacteria or prevent them from reproducing by targeting structures or biochemical processes specific to bacterial cells. Penicillins and cephalosporins disrupt synthesis of the bacterial cell wall, a rigid peptidoglycan mesh that human cells lack, causing the bacterium to burst under its own internal pressure. Fluoroquinolones like ciprofloxacin interfere with topoisomerase enzymes that bacteria use to manage their DNA. Macrolides like azithromycin bind to the bacterial ribosome, blocking protein synthesis. Each mechanism is cellular and specific.

Precision breaks down at the ecological level. An antibiotic that blocks cell wall synthesis will block it in every susceptible bacterium, regardless of whether that bacterium is causing your sinus infection or is one of the thousands of commensal species in your large intestine that ferment fiber, produce short-chain fatty acids, and keep pathogens in check. The antibiotic floods your entire body through the bloodstream, reaching every tissue including the gut lining, and acts on every susceptible organism it encounters. It has no way to distinguish a pathogen from a beneficial commensal.

Diagram 7.1. Narrow vs Broad Spectrum

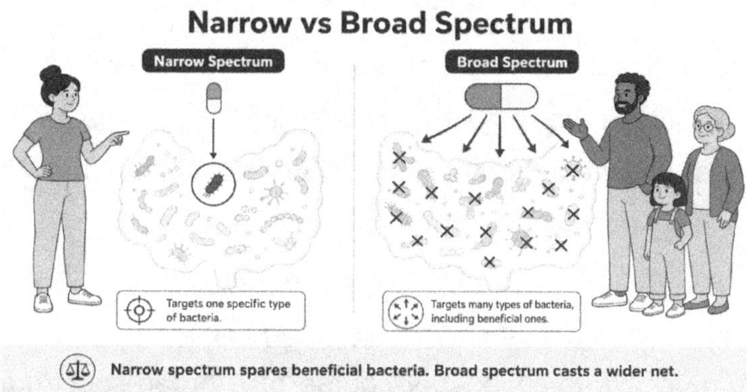

The distinction between narrow-spectrum and broad-spectrum antibiotics is crucial. Narrow-spectrum drugs target a limited range of bacterial types; broad-spectrum drugs cover many bacterial families and are often the default in primary care because they work even when the specific pathogen causing an infection has not been identified, which in routine outpatient care it usually has not.

The practical consequence is that a five-day course of a broad-spectrum antibiotic for a viral sore throat may eliminate dozens of gut bacterial species that had nothing to do with the sore throat. This is the fundamental precision problem: the weapon is blunt, the target is small, and the collateral field is enormous. When an antibiotic removes a large portion of the gut community, the species that survive are not necessarily the ones that best serve the host; they are simply the ones the antibiotic did not kill.

The Aftermath: What Happens to the Gut

The most direct way to understand how antibiotics affect the gut microbiome is to look at the data. Over the past two decades, researchers have used metagenomic sequencing to take before-and-after snapshots of gut microbial communities in people undergoing antibiotic courses, and the picture is consistently striking.

A standard five-to-seven-day course of a broad-spectrum antibiotic can reduce gut microbial species diversity by twenty-five to fifty percent within the first few days of treatment. The exact drop depends on the specific antibiotic, the individual's baseline microbiome composition, and other factors, but the direction is consistent. Some species present at low abundance before treatment are driven to functional extinction within the gut ecosystem, even if they technically persist in trace numbers. The species most vulnerable to broad-spectrum antibiotics include many that researchers associate with gut health: Bacteroides species that play key roles in polysaccharide breakdown and immune signaling, Bifidobacterium species important in early life immune development, and various Lactobacillus strains.

Diagram 7.2. Diversity Crash After Antibiotics

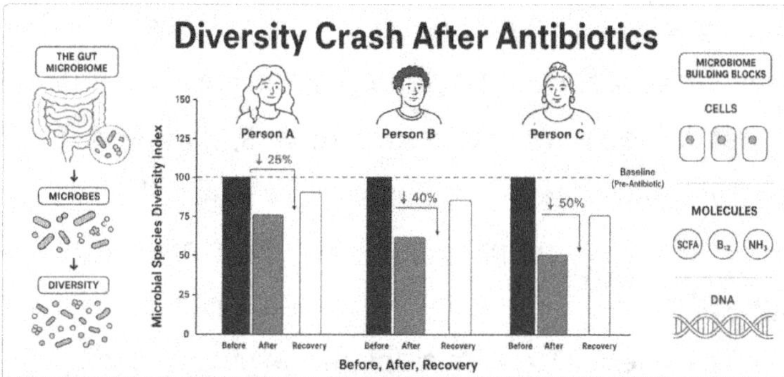

Antibiotics can significantly reduce gut microbial diversity, but partial recovery is possible over time.

What happens after the antibiotic course ends is more variable and in some ways more concerning. For many individuals, microbial diversity begins to recover within weeks, and some appear to return to a level close to their pre-antibiotic baseline within 1 to 3 months. But careful metagenomic studies that follow participants for six months to a year after treatment frequently find that, while the total species count may recover toward pre-treatment levels, the community composition differs. Species that were abundant before treatment and disappeared during it do not always return. The gut community that reassembles after antibiotics is sometimes described as a "shadow" of the original: similar in rough outline but altered in the fine-grained species relationships that determine its functional capacity.

One of the more troubling findings from metagenomic research is the expansion of the "resistome," the collection of antibiotic resistance genes in the gut microbial community. When broad-spectrum antibiotics kill the majority of susceptible bacteria, the surviving minority, which tends to carry resistance genes, is freed from competitive pressure and reproduces into the newly available space, propagating its resistance genes throughout the gut community. Repeated antibiotic courses amplify this effect. A gut microbiome exposed to multiple rounds of broad-spectrum antibiotics may harbor a substantially enriched collection of resistance genes, even in a person who has never had an antibiotic-resistant infection. This reservoir in the healthy microbiome is one of the mechanisms by which antibiotic resistance spreads through populations.

C. diff: The Opportunistic Pathogen

Clostridioides difficile provides the clearest and most clinically significant example of what happens when the protective ecology of the gut microbiome is abruptly destroyed. In healthy adults with intact, diverse gut microbiomes, C. diff is present in roughly 5% of the population at low abundance and causes no symptoms. The surrounding microbial community keeps it in check through competitive exclusion: there is not enough ecological space or available nutrients for C. diff to expand. The diverse array of bacteria occupying the gut acts as a living defensive wall.

When antibiotics demolish that wall, the situation changes dramatically. C. diff produces tough, heat-resistant spores that survive antibiotic exposure, waiting dormant while the antibiotic kills the protective community around them. When the course ends, and the gut is left as a relative ecological desert, the spores germinate. C. diff finds an environment of unprecedented abundance: colonization sites formerly occupied by protective species are empty, nutrient sources are freely available, and the immune surveillance systems that depended on microbial signals are operating in a degraded state. The result is explosive proliferation.

Diagram 7.3. C. diff Opportunity Window

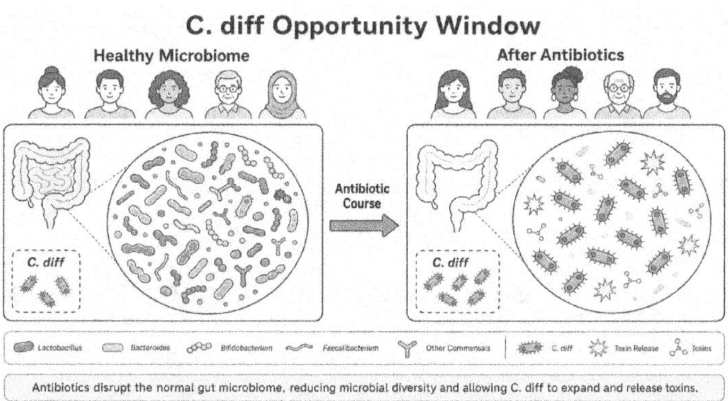

As C. diff proliferates, it releases two primary toxins that attack the colon lining. Toxin A disrupts tight junctions between intestinal epithelial cells, triggering the profuse watery diarrhea that is the hallmark of C. diff infection. Toxin B is more potent, causing direct cell death in the intestinal lining. Together, these toxins produce a spectrum of disease ranging from mild, self-limiting diarrhea to pseudomembranous colitis, to toxic megacolon, a life-threatening emergency in which the colon becomes dramatically distended and may perforate.

The cruel irony of C. diff treatment is the paradox Margaret experienced. The standard first-line treatment is another antibiotic, metronidazole or oral vancomycin, which suppresses C. diff in the short term but continues to suppress the remaining gut microbiome diversity that provides long-term protection against relapse. This is why C. diff recurrence rates after antibiotic treatment are so high: somewhere between fifteen and thirty percent of patients who recover from an initial C. diff infection experience at least one relapse, and with each relapse the probability of another increases.

The metagenomic lens fundamentally changes how we understand C. diff. For decades, C. diff was understood primarily as a problem of a dangerous pathogen: prevent exposure to C. diff spores, and you prevent disease. Metagenomics revealed that C. diff is also, in a deep ecological sense, a problem of a destroyed protective community. Many people already carry C. diff before entering a hospital. The question is not whether they encounter it; it is whether their gut community is intact enough to prevent it from taking over. This ecological framing is also the basis for fecal microbiota transplantation, the transfer of fecal matter from a healthy screened donor into the gut of the C. diff patient, which has shown cure rates of eighty to ninety percent for recurrent C. diff in clinical trials. The treatment works not by targeting C. diff directly but by restoring the ecosystem in which it cannot dominate.

Antibiotic Resistance: A Metagenomic Crisis

Antibiotic resistance has become one of the most urgent public health crises of our time. Bacteria have been producing natural antibiotic compounds and developing resistance to them for hundreds of millions of years, long before humans synthesized any antibiotics. What is new is the scale and pace of resistance development, driven by massive global antibiotic use over the past eighty years, and the increasingly sophisticated picture that metagenomics is revealing of how resistance genes spread.

Before metagenomics, the standard model of resistance focused on vertical gene transfer: random mutations arise, resistant mutants survive better in the presence of an antibiotic, and the resistant strain comes to dominate. This model is real but incomplete. What metagenomics has revealed is the pervasive importance of horizontal gene transfer, the direct sharing of genetic material between different bacteria, sometimes bacteria of entirely different species, without any reproduction involved.

Diagram 7.4. Horizontal Gene Transfer

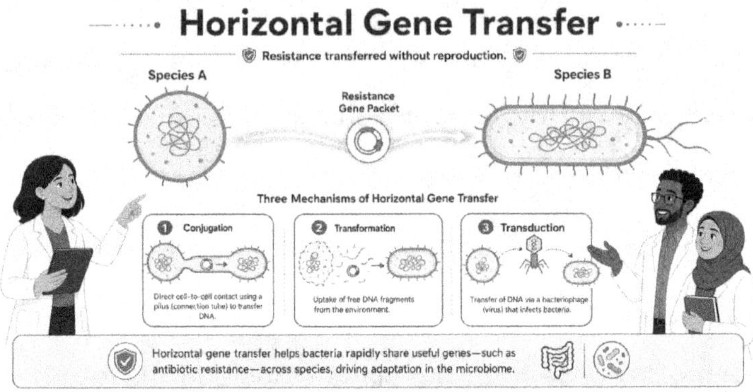

In conjugation, two bacteria form a temporary physical connection through a pilus and pass a plasmid directly between them. In transformation, a bacterium picks up naked DNA fragments released by a dead neighboring cell. In transduction, bacteriophages accidentally carry bacterial DNA from one host to another. The gut microbiome, with its extraordinary density of trillions of bacterial cells in close physical contact, is an exceptionally efficient venue for all three mechanisms, creating what researchers have described as a gene trading floor for resistance determinants.

What makes horizontal gene transfer so alarming is that resistance genes can move between bacteria from entirely different branches of the microbial tree. A resistance gene that evolved in a gut commensal species can end up in a dangerous pathogen. A gene conferring resistance in a harmless gut bacterium from a farm animal can be transferred to a human pathogen in the human gut. When researchers sequence the resistomes of gut microbiomes from people in different parts of the world, those living in settings with high antibiotic use carry significantly richer and more diverse resistomes than those from low-antibiotic-use settings.

Diagram 7.5. The Resistome Map

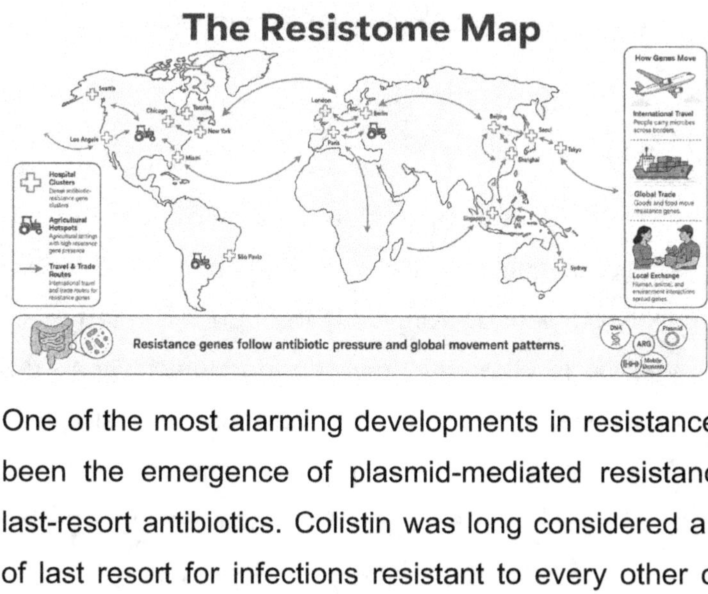

One of the most alarming developments in resistance has been the emergence of plasmid-mediated resistance to last-resort antibiotics. Colistin was long considered a drug of last resort for infections resistant to every other class. Resistance to colistin was assumed to arise only rarely through chromosomal mutation. In 2015, researchers in China identified a plasmid-borne colistin resistance gene, mcr-1, that could spread rapidly via horizontal gene transfer. Within months, mcr-1 was found in bacteria from dozens of countries, having apparently spread silently in agricultural settings for years, invisible to surveillance until metagenomics made it visible.

Understanding resistance as a metagenomic and ecological phenomenon changes the framing of the problem. Every unnecessary antibiotic use adds selection pressure to the collective resistome of the community, the hospital, the region. The consequences are distributed across people and time in ways that are genuinely difficult to perceive from inside a single clinical encounter.

Early Childhood Antibiotics: The Long Shadow

Of all the findings to emerge from the intersection of antibiotic research and microbiome science, some of the most significant concern the very beginning of life. The early childhood period, from birth through the first two or three years, is the window during which the gut microbiome establishes itself and performs its most critical developmental functions, including calibrating the immune system. What happens to the microbiome during this window can have disproportionate and long-lasting effects.

Antibiotics are the most commonly prescribed medication in children in the industrialized world. A substantial proportion of children receive their first course in the first year of life, often for ear infections, upper respiratory infections, or throat infections, many of which are viral in origin and would resolve without antibiotic treatment.

Large-scale epidemiological studies, many using national health registry data from Scandinavian countries with comprehensive record-keeping, have consistently found that cumulative antibiotic burden in early childhood is associated with significantly elevated risk of multiple chronic conditions.

Diagram 7.6. Childhood Courses Tracked

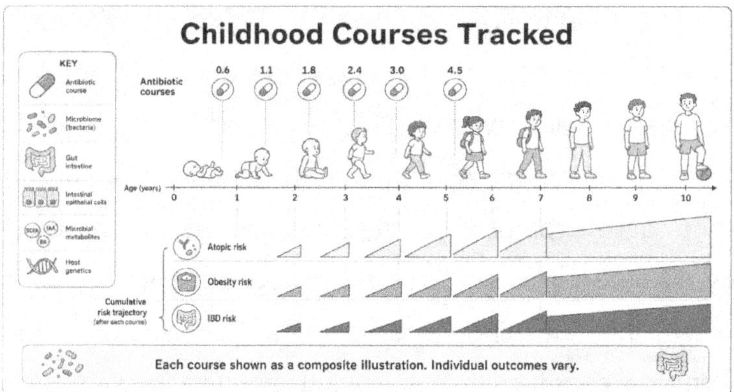

Each antibiotic course in the first two years of life is associated with increased risk of allergic sensitization and atopic conditions including asthma and eczema. The associations with obesity and overweight in later childhood are among the most replicable findings in the pediatric antibiotic literature, with the strongest effects seen when exposure occurs in the first six months of life. Associations with inflammatory bowel disease and altered metabolic parameters have been reported in multiple independent cohort studies.

The mechanism is ecological. The developing gut microbiome in early childhood is in active two-way communication with the developing immune system, which is learning to distinguish harmless environmental substances from dangerous pathogens. It receives much of this instruction through molecular signals from the microbial community: bacterial metabolites, cell wall components, and the broader pattern of microbial diversity. When antibiotics interrupt this signaling during the critical developmental window, the immune system's education is disrupted. The immune system that develops in the relative absence of microbial diversity may be more prone to misidentifying harmless substances as threats, which is precisely the defining characteristic of allergic and autoimmune conditions. The obesity associations operate through related yet distinct pathways, including disruption of the bacterial communities that produce short-chain fatty acids that regulate appetite hormones and metabolic set points.

None of this evidence means that a child who received antibiotics for a genuine infection is destined for a lifetime of health problems. The associations are statistical relationships at the population level, and individual outcomes depend on a dense network of genetic, environmental, dietary, and socioeconomic factors. The appropriate response is not to withhold antibiotics from children who genuinely need them; it is to approach antibiotic prescribing for children with careful proportionality, recognizing that unnecessary courses carry real downstream costs.

Microbiome Recovery: Can It Bounce Back?

One of the most common questions people ask after learning about antibiotic effects on the gut microbiome is whether the microbiome recovers after a course ends. The answer is: it depends, in ways that are increasingly well understood and practically informative.

The gut microbiome does have significant resilience. Bacterial communities in the gut have always been subject to perturbations, including transient pathogen invasions, dietary changes, and illness. The capacity to recover from disturbance is itself an evolved feature of the microbial ecosystem. For many healthy adults who receive a single course of a standard antibiotic, measurable recovery in microbial diversity begins within days of completing the course. It may reach levels close to pre-treatment within 1 to 3 months.

But "something close" is not the same as "identical." The most careful metagenomic studies, those following the same individuals for 6 months to a year post-treatment, consistently find that the composition of the restored community differs from that of the pre-treatment community. Some species eliminated during treatment never return within the observation window. Others return at altered abundances. The recovered microbiome may be functionally similar in many respects while differing in specific capacities, such as the ability to produce particular metabolites or respond to specific immune signals.

Diagram 7.7. Recovery Curves: Fast vs Slow Bouncers

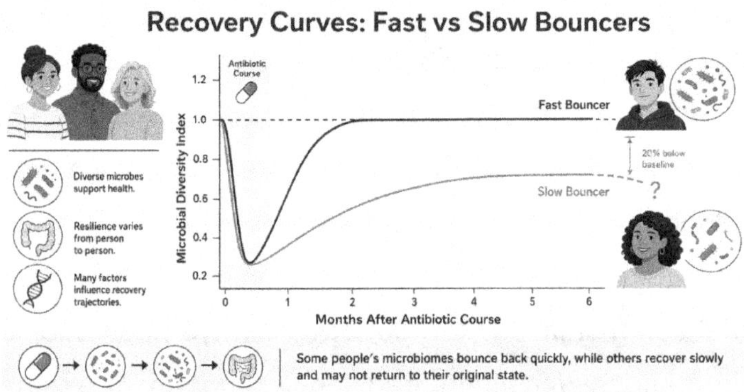

The rate and completeness of recovery vary substantially between individuals. Baseline microbial diversity before treatment is among the most important factors: people who start with high diversity tend to recover more completely because more species are available to recolonize niches left by antibiotic-induced clearance. The type of antibiotic matters significantly: fluoroquinolones and clindamycin tend to produce more severe and prolonged disruptions than narrow-spectrum beta-lactams. The number of prior antibiotic courses has a cumulative effect; each successive course tends to reduce baseline diversity, making recovery from subsequent courses progressively less complete.

One of the most counterintuitive findings in the recovery literature concerns probiotics. Conventional wisdom holds that taking probiotics during and after antibiotic use helps restore the gut microbiome. A series of carefully designed studies, including work from Eran Segal's laboratory in Israel, found that this intuition is not quite right. Participants randomized after antibiotic treatment to receive a standard commercial probiotic product containing eleven bacterial strains recovered their pre-antibiotic microbiome composition significantly more slowly than participants who received no treatment at all. The probiotic strains were occupying ecological space that would otherwise have been available to the returning native microbiome. The fastest recovery occurred in a group that received fecal microbiota transplantation using their own pre-antibiotic stool, essentially having their original microbiome restored directly.

This does not mean probiotic products are without value in all contexts. Specific strains have been shown in well-designed trials to reduce antibiotic-associated diarrhea, the general loose stools many people experience during treatment. Lactobacillus rhamnosus GG and Saccharomyces boulardii (a beneficial yeast) have the strongest evidence for this specific purpose. But for restoring the microbiome to pre-antibiotic composition, off-the-shelf commercial probiotic products are at best neutral and potentially counterproductive. In contrast, the microbiome's natural recovery capacity, supported by a fiber-rich diet, is the more reliable path.

Antibiotic Stewardship: The Path Forward

The evidence in this chapter builds toward a practical conclusion: antibiotics are indispensable, but the way they have been and continue to be prescribed in many healthcare settings is not consistent with what we now know about their ecological consequences. Changing that is the goal of antibiotic stewardship, the systematic, evidence-based management of antibiotic prescribing to ensure these drugs are used only when genuinely needed, with the narrowest appropriate spectrum, at the correct dose, and for the correct duration.

In hospitals, stewardship programs typically involve pharmacy and infectious disease specialists reviewing prescriptions in real time, ensuring that prescribed antibiotics match culture results when available, tracking resistance patterns, and educating clinical staff. Multiple studies have shown that well-implemented hospital stewardship programs reduce C. diff infection rates, reduce the emergence of antibiotic-resistant organisms, and reduce costs, all without any measurable increase in patient harm.

In primary care settings, where the majority of antibiotic prescriptions are written, stewardship operates during brief clinical encounters, under time pressure, without culture data in most cases, and amid strong patient expectations. Changing those cultural expectations requires communication, education, and a shared understanding between patients and clinicians about the genuine costs and benefits of antibiotic treatment.

One practical tool from stewardship research is the "wait-and-see" or "delayed" prescription: the clinician prescribes the visit but asks the patient to wait two or three days before filling it, doing so only if symptoms do not improve. Studies find that the majority of patients who receive a delayed prescription for ear infections, mild respiratory infections, and sore throat never fill it because the condition resolves on its own.

Rapid diagnostic tests are another emerging stewardship tool. One of the most common scenarios for antibiotic overprescribing is a sore throat with fever, which can be caused by group A Streptococcus, for which antibiotics are beneficial, or by a range of viruses, for which antibiotics are ineffective. A rapid streptococcal antigen test provides a result in minutes, allowing targeted prescribing. The test is inexpensive and widely available but not universally used. Expanding its use is one of the clearest low-effort gains in outpatient stewardship.

The narrow-spectrum-first principle is one that stewardship advocates consistently emphasize. If a patient has a skin infection whose clinical presentation suggests it is gram-positive, a narrow-spectrum antibiotic is more appropriate than a broad-spectrum antibiotic. If a urinary tract infection has known susceptibility to trimethoprim, there is no microbiome-protective reason to use a fluoroquinolone. In every clinical situation where a narrower antibiotic will work as well as a broader one, the narrower choice is preferable for the patient's microbiome and the community's collective resistome.

For individual readers, the most direct entry point into the stewardship conversation is the clinical encounter where an antibiotic is being considered. The questions in the diagram below are starting points for those conversations, designed to ensure the decision is made with full information and genuine consideration of alternatives.

Diagram 7.8. Questions to Ask Your Doctor

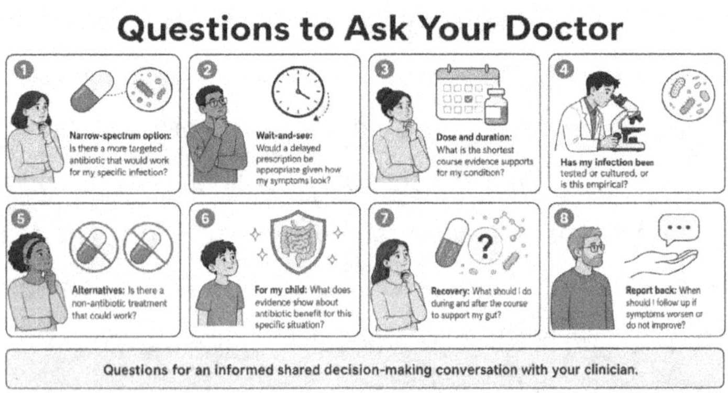

Antibiotic stewardship is ultimately a shared responsibility operating at every level from international regulation to the individual clinical encounter. The half-million Americans who develop C. diff infections each year, the millions of children whose microbiomes are disrupted by unnecessary prescriptions, and the patients around the world who will need antibiotics in twenty years and may find the drugs no longer work, are all part of the same story. It is a story that metagenomics is telling with unprecedented clarity, and one that both clinicians and patients now have the information to help change.

Antibiotic stewardship is, at its core, an ecological argument made urgent by clinical stakes. Of the practices stewardship programs advocate, the one that matters most in everyday practice is the narrow-spectrum-first principle — because it is the decision point that clinicians face most often, on the most patients, with the clearest evidence for an alternative. When trimethoprim will work as well as a fluoroquinolone for a urinary tract infection, there is no microbiome-protective reason to choose the broader drug. That single habit, applied consistently in outpatient settings, would meaningfully reduce the ecological damage done per prescription year.

The wait-and-see prescription is the second practice worth highlighting, not because it is the most intellectually interesting but because it works, is inexpensive, and requires nothing more than a conversation. Most ear infections, most sore throats, most mild respiratory illnesses will resolve on their own. Giving patients a prescription to fill only if symptoms worsen after two days captures this reality without abandoning them in an exam room empty-handed. Studies consistently find that most patients never fill it.

The half-million Americans who develop C. diff infections each year and the millions of children whose microbiomes are disrupted by unnecessary prescriptions are part of the same ecological story that metagenomics is now able to tell with unusual clarity. Clinicians and patients together have the information needed to change its ending. That is an unusually hopeful sentence to write about a public health problem of this scale.

Takeaways

Understanding the antibiotic paradox gives readers a framework for using one of medicine's most important tools more wisely, without fear and without avoidance, but with clear-eyed awareness of both its power and its ecological costs.

- Antibiotics are not precision tools at the ecological level. A five-day broad-spectrum course can reduce gut microbial species diversity by twenty-five to fifty percent, and some of that diversity may not fully return for months or at all.

- C. diff is a microbiome story, not just an infection story. C. diff causes disease because antibiotics destroy the protective gut community that normally keeps it in check. The treatment paradox (using more antibiotics to fight an antibiotic-caused infection) is real, which is why fecal microbiota transplantation now achieves far higher cure rates for recurrent C. diff than antibiotics alone.
- Antibiotic resistance spreads through the gut as an ecological and genetic phenomenon. Resistance genes move between bacterial species through horizontal gene transfer. Every unnecessary antibiotic course enriches the local resistome and contributes to a global public health crisis already killing hundreds of thousands annually.
- Early childhood antibiotic exposure carries disproportionate risk. The gut microbiome in the first two years of life is calibrating the immune system. Disruptions during this window are associated at the population level with elevated risk of atopic disease, obesity, and inflammatory bowel disease. Antibiotics remain appropriate for genuine infections in children; the goal is to reduce unnecessary courses.
- Microbiome recovery is possible but not guaranteed to be complete. Most healthy adults recover substantial microbial diversity within one to three months after a

single antibiotic course. A high-fiber diet supports recovery. Counterintuitively, standard commercial probiotic products may slow natural recovery by occupying ecological space that native bacteria would otherwise reclaim.

- Antibiotic stewardship is evidence-based medicine, not antibiotic avoidance. Stewardship programs reduce C. diff rates, the emergence of resistance, and costs without any measurable increase in patient harm. At the individual level, stewardship means asking informed questions before accepting a prescription, preferring narrow-spectrum options when clinically adequate, and using delayed prescriptions when appropriate.
- Before your next antibiotic prescription, consider asking: Is there a narrow-spectrum antibiotic that would work for my specific infection? Would a wait-and-see approach be appropriate? What is the shortest course of evidence that supports my condition? Has my infection been tested or cultured? Are there non-antibiotic alternatives? And if a child is involved: what does the evidence actually show about antibiotic benefit for this specific diagnosis in this age group?

8 Engineering the Microbiome

The waiting room outside the gastroenterology clinic at the University of Minnesota had the particular stillness of a place where people arrive with problems that polite conversation cannot fix. In the autumn of 2012, a woman in her late fifties sat in one of those chairs having already lost count of the antibiotic courses behind her. Two years of recurrent Clostridioides difficile infection had stripped her gut of nearly everything healthy. Each round of antibiotics would subdue the infection for a few weeks, only for C. diff to bloom again, exploiting the ecological vacuum left behind. She had lost weight she could not afford to lose. She had stopped going out. She had organized her life around the distance to the nearest bathroom.

Her gastroenterologist, Alexander Khoruts, was considering an older idea that most colleagues still found distasteful. If a gut microbiome is devastated, why not replace it the same way you might replace a failing organ? A healthy donor, rigorously screened for pathogens, would provide a stool sample. The sample would be processed, filtered, and delivered to the recipient's colon, either via colonoscope or swallowed capsule. The donor's microbial community would take up residence and, if things went as hoped, permanently crowd out C. diff.

Within twenty-four hours of her fecal microbiota transplant, the woman's diarrhea had stopped. Within three months, metagenomic sequencing showed something extraordinary: her original microbiome, battered into near-extinction by two years of antibiotics and C. diff cycles, was essentially gone. What remained looked like her donor's microbiome. She had not just been treated; she had been, in a meaningful ecological sense, repopulated. Khoruts published the case. Then came the randomized trials, the replications, the regulatory review, and eventually, in 2022 and 2023, FDA approval of the first manufactured microbiome-based drugs for recurrent C. diff. A fringe idea had become mainstream medicine, and it had done so not through hype but through rigorous evidence. That story is the spine of this chapter.

Consider another reader: Harold, sixty-seven, a retired schoolteacher in Bend, Oregon. He has just read a Sunday newspaper article about fecal microbiota transplantation and is sitting with his coffee, half-amused and half-incredulous. The article mentioned an FDA-approved version and quoted a physician using the phrase 'standard of care' for recurrent C. difficile. Harold has no C. difficile. He has no acute reason to be reading about this at all. What he has is forty-one years of teaching biology to ninth graders, a habit of asking questions until something becomes clear, and a granddaughter in medical school whom he would like to be able to talk to about it. He is the reader this chapter is also for: not a patient looking for a cure but a citizen trying to understand a real piece of medicine that has moved from fringe to mainstream in his lifetime.

Why Intervention Matters

Most of us encounter the microbiome through the lens of intervention, even if we do not use that word. We read about probiotics on a yogurt label. We hear about prebiotics from a nutrition podcast. We scroll past an advertisement for a gut-health supplement that, in carefully hedged language, claims to support digestive wellness. And now, we increasingly see news about fecal transplants, bacteriophage therapy, and live biotherapeutic products advancing through clinical trials. The microbiome has become one of the most commercially exploited territories in health science, which makes it one of the most confusing.

This is also where the earlier virome discussion comes back into focus. Phage therapy is not an exotic side branch of microbiome science; it is a practical extension of the idea that viruses, especially bacteriophages, are active members of microbial ecosystems and can sometimes be directed with precision against problematic bacterial strains when broad-spectrum antibiotics create more collateral damage than benefit.

This chapter walks through every major category of microbiome intervention currently available or in serious clinical development, examines what the evidence actually shows, and offers a framework for distinguishing legitimate science from marketing dressed up in scientific language. The goal is not skepticism; it is precision. Some interventions are genuinely impressive, and the frontier science is genuinely exciting. The goal is to help you ask the right questions before swallowing a pill, eating a specific food, or consenting to a medical procedure that involves another person's microbiome.

Consider a composite story drawn from many real patients and caregivers: Robert, a sixty-year-old accountant in suburban Maryland, has spent eight months helping his eighty-three-year-old mother, Eleanor, navigate recurrent C. diff infections following hip replacement surgery. Eleanor had her third episode six weeks ago. She is tired, thin, and frightened. Robert has been researching her options and has come across fecal microbiota transplantation. He is not sure whether it is genuine medicine or an internet curiosity. He has questions about probiotics, about diet, about whether there is anything he can do for her while they wait for the next specialist appointment. Keep Robert and Eleanor in mind as you read; the framework that helps them is the same one that will help you.

What Probiotics Actually Are and Are Not

The word "probiotic" is one of the most commercially productive and scientifically abused terms in consumer health. Walk through any pharmacy or natural foods store, and you will encounter hundreds of products labeled as probiotics, promising everything from improved digestion to better mood to enhanced immune function. Very few of those claims are supported by the kind of rigorous clinical evidence that should make you reach for your wallet.

The technical definition, established by a joint expert panel of the Food and Agriculture Organization and World Health Organization, is precise: a probiotic is a live microorganism that, when administered in adequate amounts, confers a health benefit on the host—every word matters. "Live" means the organism must survive manufacture, packaging, and storage to reach your gut still viable. "Adequate amounts" means the dose must be high enough, typically measured in colony-forming units (CFUs), and that number must hold through the shelf life. "Confers a health benefit" means the benefit must be demonstrated in rigorous human studies, not inferred from cell cultures or animal models.

Most commercially available probiotic supplements contain strains from two genera: Lactobacillus and Bifidobacterium. These organisms have genuine evidence supporting specific, narrow applications. The evidence for probiotics is profoundly strain-specific. Lactobacillus rhamnosus GG (LGG) has reasonably strong evidence supporting its reduction of antibiotic-associated diarrhea in children. That evidence does not transfer to other Lactobacillus strains, even closely related ones. The genus name on the label tells you almost nothing about whether the specific organism inside will do what the marketing promises.

Diagram 8.1. The Three Bios

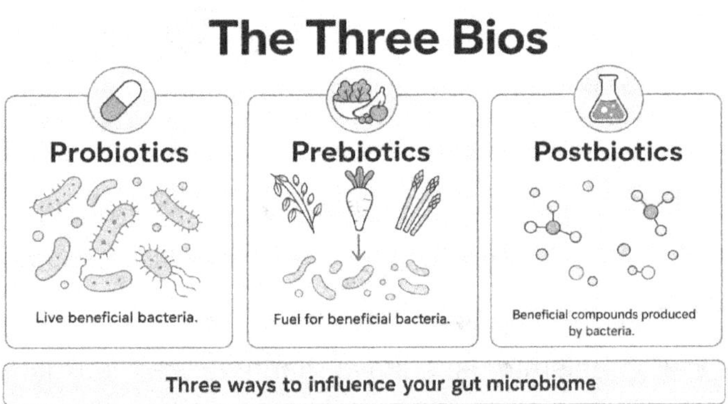

When researchers apply the scrutiny of systematic reviews and meta-analyses to probiotic claims, what emerges is modest, conditional benefit rather than sweeping health transformation. For antibiotic-associated diarrhea, the evidence is genuinely encouraging: well-conducted randomized controlled trials show that LGG or Saccharomyces boulardii (a beneficial yeast) taken alongside antibiotics reduces the risk and duration of diarrhea. For irritable bowel syndrome, certain strains appear to reduce symptom scores in some trials. Still, the effects are inconsistent, and it is not yet possible to identify which strain to recommend for which IBS patient. For general immune support, weight management, and skin conditions in healthy adults, the evidence is thin to nonexistent.

There is also a practical problem most consumers never consider: survival. Your stomach, with a pH typically between 1.5 and 3.5, kills a meaningful fraction of any probiotic dose before it reaches the colon. Manufacturers address this with enteric coating and high initial CFU counts. Still, enteric coating quality varies widely, and independent testing has repeatedly shown that many supplements do not contain the viable organism count stated on the label. That listed strains do not always match those present. In the United States, probiotic supplements are regulated as dietary supplements and do not require FDA approval before going to market.

Diagram 8.2. The Probiotic Pill Journey

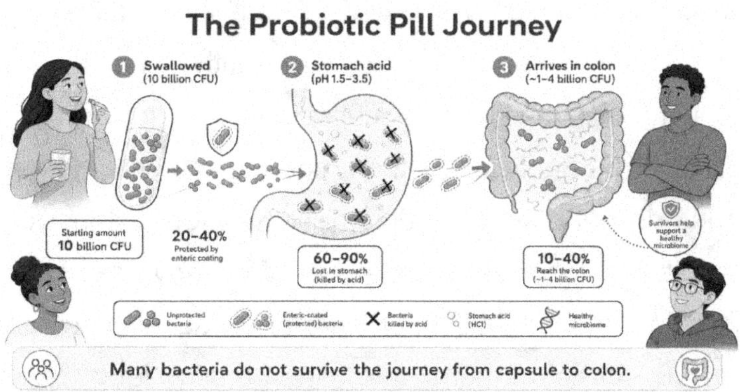

None of this means probiotics are useless. It means you should take them strategically. If you are taking an antibiotic and want to reduce the risk of diarrhea, reaching for LGG or S. boulardii is a reasonable, evidence-based choice. If you are a healthy adult taking a daily probiotic supplement hoping for general health improvements, the evidence does not strongly support that expectation. And if your primary concern is improving the overall composition and diversity of your gut microbiome, the most powerful lever is not a capsule; it is what you eat.

Prebiotics: Feeding What's Already There

The metaphor of a garden is useful here. Probiotics are like planting new seeds; prebiotics are like improving the soil for what is already growing. The prebiotic approach works with the microbial residents already present, providing them with the specific nutrients they need to thrive, reproduce, and out-compete less beneficial species. A prebiotic is technically a substrate selectively utilized by host microorganisms to confer a health benefit. The "selectively utilized" part is critical: it has to be food specifically for beneficial bacteria, not any bacteria. Human enzymes do not digest the dietary fibers most extensively studied as prebiotics in the small intestine; they pass intact to the colon, where Bifidobacterium and Faecalibacterium prausnitzii ferment them into short-chain fatty acids (SCFAs). Butyrate, the most important SCFA, is the primary energy source for colonocytes (the cells lining your colon) and has anti-inflammatory properties that extend well beyond the gut.

The most thoroughly studied prebiotic compounds include inulin and fructooligosaccharides (FOS), found in chicory root, garlic, onions, and Jerusalem artichoke; beta-glucans, found in oats and barley; and resistant starch, found in cooked-and-cooled potatoes, green bananas, and legumes. Each travels largely intact to the colon. Feeding trials consistently show increases in Bifidobacterium and SCFA production. The mechanistic logic is solid: more butyrate means better gut-lining integrity, lower colonic inflammation, and a more hospitable environment for diverse microbial life.

Diagram 8.3. Prebiotic Foods

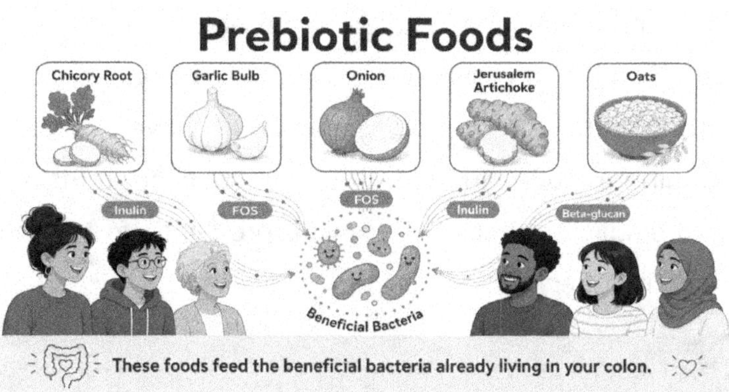

What makes the prebiotic evidence compelling is that it works with the ecology already present rather than introducing a foreign organism. Diverse plant fiber intake is consistently associated with greater gut microbial diversity, higher SCFA production, and reduced systemic inflammation across observational studies and intervention trials. The approach is primarily dietary: eat more diverse plants, with particular emphasis on fibrous vegetables, legumes, whole grains, and alliums (garlic, onion, leek, chive family).

Prebiotic supplements are not without merit. For people who genuinely cannot eat sufficient dietary fiber, inulin powder, FOS capsules, or beta-glucan supplements offer a practical alternative. Their main side effect at high doses, gas and bloating, is itself evidence that the fiber is reaching the colon and being fermented. Synbiotics, combining probiotics and prebiotics in a single product, aim to introduce beneficial organisms while simultaneously providing them with the nutrient substrates they need to survive. Studies on synbiotics for antibiotic-associated diarrhea and metabolic syndrome have shown encouraging results, though the key question remains: which probiotic, paired with which prebiotic, for which condition, in which patient?

Postbiotics: The Next Frontier

If probiotics are the live performers and prebiotics are the stage, postbiotics are the music itself. They are the bioactive compounds that gut bacteria produce, the molecular signals and substrates through which the microbiome communicates with the rest of the body and delivers its health effects. Rather than adding live organisms or feeding them, a postbiotic product delivers these compounds directly, bypassing the need for bacteria to produce them in situ.

The category is formally defined by the International Scientific Association for Probiotics and Prebiotics as a "preparation of inanimate microorganisms and/or their components that confers a health benefit on the host." The definition is deliberately broad. It encompasses short-chain fatty acids (particularly butyrate, propionate, and acetate), exopolysaccharides produced by bacteria, bacterial cell wall components like peptidoglycan fragments, bacteriocins (natural antibiotics produced by bacteria), and heat-killed bacteria that retain biological activity.

Butyrate is the most well-characterized postbiotic. As the primary fuel for colonocytes, it maintains gut-lining integrity, reduces intestinal permeability (colloquially called "leaky gut"), and suppresses colonic inflammation. Butyrate supplements have been studied in inflammatory bowel disease and irritable bowel syndrome. The challenge is that orally ingested butyrate is absorbed primarily in the small intestine and may not reach the colon in effective concentrations; enteric-coated formulations designed to release butyrate in the colon are being developed to close that gap.

The broader appeal of postbiotics is practical: they are shelf-stable, can be manufactured with consistent composition, and are safe for immunocompromised patients who might react poorly to live bacteria. Most postbiotic products on the market are ahead of the clinical evidence supporting them, and consumers need to carefully distinguish products backed by genuine research from those borrowing the postbiotic label for marketing purposes.

Fecal Microbiota Transplantation: The Evidence

Return to Robert and Eleanor for a moment. After reading everything he can find about Eleanor's recurrent C. diff, Robert has landed on fecal microbiota transplantation as the option most discussed in credible sources. He is simultaneously convinced that this might be exactly what Eleanor needs and unsettled by the very concept. He wonders whether it is real medicine or alternative medicine with a clinical-sounding name. He wonders where the donor comes from, how the procedure is done, whether it is safe for an eighty-three-year-old, and whether Eleanor's doctor will take his research seriously.

Robert's questions are exactly the right ones, and the answers in Eleanor's case are substantially reassuring. FMT is the most rigorously evidence-based microbiome intervention. Multiple randomized controlled trials conducted in different countries by independent research groups have consistently shown FMT cure rates of 80 to 95 percent for patients with recurrent C. diff, compared with 30 to 65 percent for the best available antibiotic regimens. That is a clinically meaningful and statistically robust difference for a condition that can be life-altering.

The mechanistic rationale is straightforward. Recurrent C. diff is, at its core, an ecological problem. C. diff flourishes in a gut whose normal microbial community has been disrupted, typically by antibiotics. It produces toxins that damage the gut lining and cause severe watery diarrhea. Every antibiotic course used to treat it further depletes the surrounding microbial community, creating more ecological space for C. diff to rebound when the antibiotics end. FMT breaks the cycle by restoring the ecology, populating the gut with a dense, diverse consortium from a healthy donor that can resist C. diff colonization, produce the short-chain fatty acids and secondary bile acids that suppress C. diff germination, and restore the integrity of the gut lining.

Diagram 8.4. The FMT Process

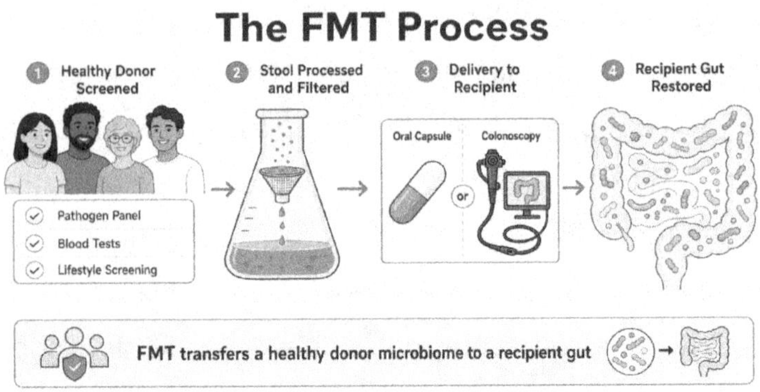

The delivery method has evolved considerably. The original approach was colonoscopic delivery, administering processed donor material directly to the colon via colonoscope. That remains common and allows direct visualization. More recently, encapsulated FMT, in which frozen donor material is placed in acid-resistant capsules and swallowed, has been shown in randomized trials to be non-inferior to colonoscopic delivery, with the practical advantage of not requiring a procedural suite, sedation, or bowel preparation.

The regulatory landscape shifted decisively in 2022 and 2023. The FDA approved Rebyota (Ferring Pharmaceuticals), a rectally administered microbiome product derived from screened donor stool, for prevention of recurrent C. diff in adults. In 2023, the FDA approved Vowst (Seres Therapeutics), an oral capsule containing spores of donor-derived bacteria, for the same indication. These approvals were a watershed moment: for the first time, a manufactured product containing donor microbiome material had completed the full FDA approval pathway, demonstrating safety, efficacy, and manufacturing consistency. The era of microbiome medicine had officially begun.

The safety profile of FMT for C. diff, when performed by experienced practitioners using rigorously screened donors, is favorable. The most important concern is the transmission of pathogens from donor to recipient, which is managed through comprehensive screening. Manufactured products such as Rebyota and Vowst undergo industrial-scale quality controls in every production run. Beyond C. diff, the FMT evidence base is much more preliminary; trials for ulcerative colitis have shown mixed results. For Eleanor and Robert, the message is clear: for recurrent C. diff, FMT is not experimental medicine; it is the standard of care.

Phage Therapy: Precision Strikes Against Bacteria

Bacteriophages are viruses that infect bacteria, and only bacteria; they cannot infect human cells. Phages are the most numerous biological entities on Earth, with an estimated 10^31 individual particles in the oceans alone. Every species of bacterium has natural phage predators that have evolved alongside it over billions of years, and those relationships are exquisitely specific. A phage that infects Escherichia coli typically cannot infect

Staphylococcus aureus. This specificity is simultaneously the most exciting property of phage therapy and its greatest practical challenge.

The idea of using phages therapeutically predates the antibiotic era. Felix d'Herelle, who co-discovered bacteriophages in the 1910s, was an early proponent. Phage therapy was practiced widely in France, Georgia, and Poland through the mid-twentieth century, and the George Eliava Institute in Tbilisi still operates today. In the West, phage therapy was displaced by antibiotics after World War II; antibiotics were cheaper and easier without the need to identify a specific phage for each pathogen. The antibiotic resistance crisis has revived serious interest among Western policymakers with considerable urgency.

The appeal of phage therapy is threefold. Phages are self-replicating: a small dose can multiply at the site of infection. They can be highly specific, targeting a pathogen while leaving the beneficial microbial community intact. And phage cocktails can be engineered to reduce the likelihood of bacterial resistance. For conditions involving recurrent or antibiotic-resistant gut pathogens, the prospect of precision phage targeting is genuinely appealing.

Diagram 8.5. Phage Therapy Precision

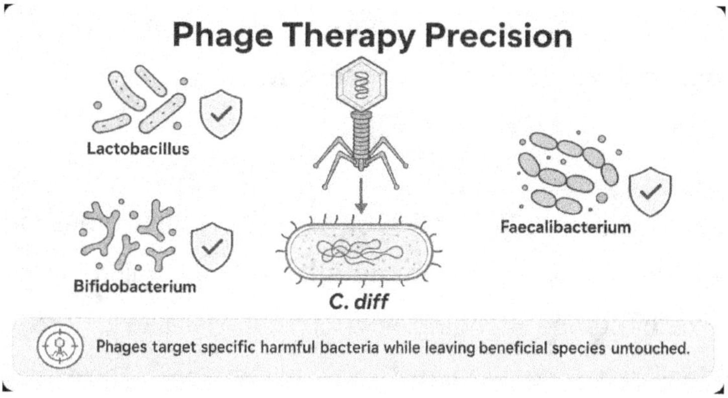

The most dramatic modern phage therapy cases have come from compassionate-use applications. Isabelle Holdaway, a fifteen-year-old in the United Kingdom with a disseminated Mycobacterium abscessus infection following a lung transplant, drew international attention in 2017: researchers identified and engineered three phages targeting her specific strain and treated her intravenously and topically. She responded. Her case and others like it demonstrated that targeted phage therapy is not merely theoretically possible but practically achievable.

For gut applications, phage therapy faces additional complexity. Targeting a specific gut pathogen requires a phage that can survive the acidic stomach, reach the colon, identify its specific target among trillions of other organisms, and replicate sufficiently to exert a meaningful effect. Early phase clinical trials are underway, and phage biobanks cataloging the activity profiles of thousands of phage strains are being built to accelerate clinical use. The field is moving, but it will be years before phage therapy is available outside of compassionate-use programs.

Designer Microbiomes and Live Biotherapeutics

Reading the word "probiotic" on a yogurt cup gives you one picture of what it means to influence the microbiome with live bacteria. Standing in a clinical trial site where a manufactured consortium of twenty-five precisely characterized bacterial strains is being administered to a patient with ulcerative colitis gives you a completely different one. The gap between those two images describes the distance between today's microbiome supplement market and what researchers call live biotherapeutic products, or LBPs.

An LBP, as defined by the FDA, is a biological product that contains living organisms and is intended to prevent, treat, or cure a disease or condition. LBPs are subject to the same rigorous preclinical testing, phase I, II, and III clinical trial requirements, and regulatory review as any other biological drug. The probiotic supplements on your pharmacy shelf operate under a completely different legal framework, one that does not require proof of efficacy before marketing. LBPs must prove they work, demonstrate they are safe, and be manufactured with pharmaceutical-grade consistency.

Several companies are now advancing LBPs through the clinical pipeline. Seres Therapeutics, the maker of Vowst, has an LBP pipeline in development for inflammatory bowel disease. Vedanta Biosciences is conducting trials of defined bacterial consortia for conditions including food allergy, IBD, and cancer immunotherapy potentiation. Synlogic is taking an even more engineered approach, creating genetically modified bacteria to perform specific therapeutic functions in the gut, such as converting ammonia in patients with urea cycle disorders or producing anti-inflammatory molecules in patients with IBD.

Diagram 8.6. Live Biotherapeutic Product

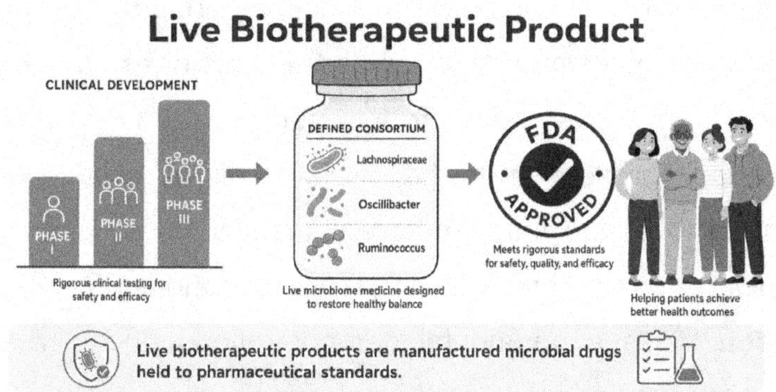

Modern LBP development involves metagenomics-guided discovery, in which researchers identify organisms consistently associated with health in large population studies or consistently absent in people with specific diseases; followed by isolation and characterization of candidates; followed by preclinical testing in organoid models; followed by the standard phases of human clinical trials. The manufacturing challenge is formidable:

producing clinical-grade quantities of obligate anaerobes requires specialized fermentation infrastructure, and quality control for living organisms is inherently more complex than for small-molecule drugs.

Precision microbiome medicine takes LBP development one step further. Rather than designing a single LBP for a broad population, researchers are exploring whether specific microbiome profiles can predict who will respond to which intervention. Early results in oncology, where microbiome composition appears to influence immunotherapy response, are compelling enough to drive substantial investment. Engineered microbes whose genomes have been edited to introduce novel therapeutic functions represent a further frontier: living medicines designed to sense inflammatory signals and respond by producing targeted anti-inflammatory compounds. The gap between these experimental demonstrations and clinical availability is significant, but the scientific basis is solid.

Separating Hype from Hope: A Practical Guide

The microbiome supplement market generates billions of dollars annually, and much of that revenue rests on claims that range from genuinely evidence-supported to entirely speculative to actively misleading. Navigating that market requires a framework, a set of questions you can apply systematically to any microbiome product or health claim. The framework is not about cynicism; it is about calibration.

The first question is the most fundamental: has this specific strain or intervention been tested in humans? The microbiome field generates enormous quantities of fascinating research in mice, in cell culture, and using computational models. That is real science, but it is not evidence that a product will work in you. The gap between a mouse study showing a bacterial species reduces inflammation in a rodent model and a supplement claiming to reduce inflammation in humans is vast, and most products bridging that gap with marketing language have not done the work to cross it legitimately.

The second question is about study design: is the evidence from randomized controlled trials? Observational studies and testimonials play a role in generating hypotheses; they are not sufficient evidence that an intervention works. The gold standard for establishing causation is the randomized controlled trial (RCT), in which participants are randomly assigned to receive either the intervention or a placebo, and neither group knows the assignment until the data are analyzed. For any microbiome product claiming a health benefit, the question is whether that benefit has been demonstrated in at least one well-conducted RCT with adequate statistical power.

Diagram 8.7. The Evidence Stratification

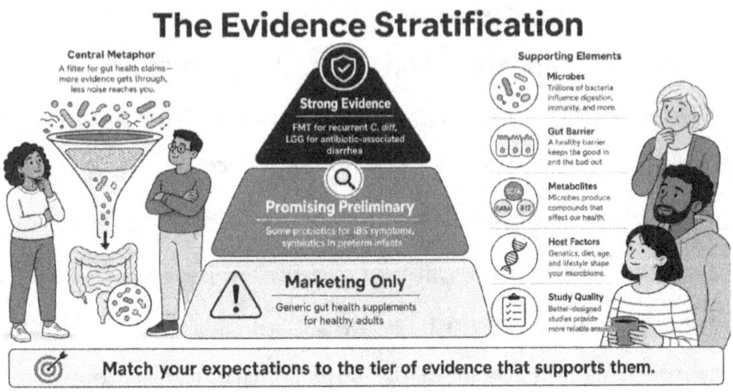

The third question is about replication: has the effect been reproduced by independent research groups? A single positive RCT is not definitive. The history of science is full of effects that seemed real in one study and shrank or disappeared when other researchers tried to reproduce them. When multiple independent groups, with no financial stake in the outcome, consistently find the same effect, confidence grows substantially. When a product's evidence base consists primarily of studies funded by the manufacturer, healthy skepticism is warranted.

The fourth question is about mechanism: is there a biologically plausible reason why this product would do what it claims? A claim that a specific probiotic strain reduces gut inflammation has a coherent mechanistic story: bacteria interact with gut immune cells, influence cytokine production, and compete with inflammatory species. A claim that the same strain improves joint pain, enhances athletic performance, and sharpens cognitive function in healthy adults requires a much longer and more speculative chain of reasoning.

The fifth question is about conflicts of interest: is the company primarily citing its own proprietary research? This is about incentive structures, not bad faith. A company that has invested in a specific strain has a financial interest in studies showing positive results for that strain, and the scientific literature reflects the well-documented tendency to publish positive results. Independent replication by researchers with no financial stake in the outcome is the closest thing we have to a truth test in the messy business of clinical research.

Diagram 8.8. Five Questions to Ask Before Buying

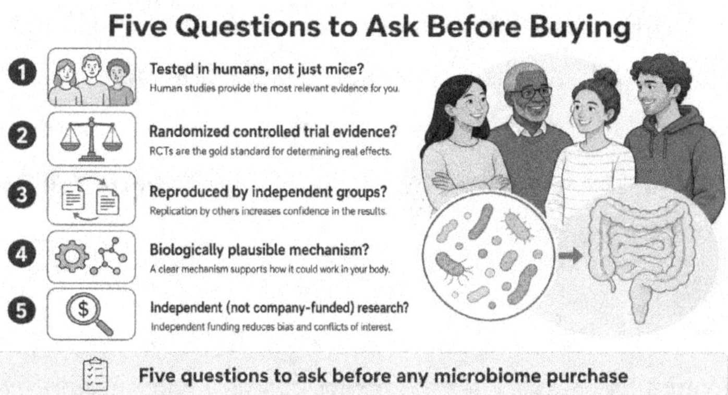

For Robert and Eleanor, this framework leads to a clear conclusion: FMT for recurrent C. diff clears every bar. The evidence comes from hundreds of randomized trials. The effect has been replicated across continents by independent research groups. The mechanism is biologically coherent and experimentally supported. The evidence base is dominated by academic investigators without financial stakes in particular products. FDA-approved manufactured versions now exist. If Eleanor's doctor agrees she is a candidate, FMT is not a leap of faith; it is a medically supported decision.

For the more typical microbiome-product decision, whether to buy a probiotic supplement, try a prebiotic fiber product, or experiment with a postbiotic preparation, the framework does not give a blanket yes or no. It gives a way to ask better questions and arrive at more grounded expectations. Some products will pass all five filters for specific applications. Many will pass; only some are worth trying at modest cost and low risk, but should not carry outsized expectations. And some will fail most filters, suggesting that the money might be better spent on vegetables and legumes that no marketing department has ever paid to promote.

Takeaways

- Probiotics are live microorganisms with evidence-based benefits for specific conditions, particularly antibiotic-associated diarrhea; the evidence does not support general gut-health claims for healthy adults, and strain specificity matters enormously.

- Prebiotics nourish the microbiome already present; diverse plant fiber intake, especially from alliums, legumes, chicory, oats, and resistant starch, is the most accessible and evidence-backed way to improve gut microbial diversity and short-chain fatty acid production.
- Postbiotics, the bioactive compounds produced by gut bacteria, are a genuinely promising frontier; the category is real and mechanistically coherent, but most commercial products are ahead of the clinical evidence that would justify specific health claims.
- Fecal microbiota transplantation for recurrent Clostridioides difficile infection is the most powerfully evidence-supported microbiome intervention available, with cure rates of 80 to 95 percent in randomized trials; FDA-approved manufactured products now make FMT legitimate and accessible medicine.
- Phage therapy offers the possibility of eliminating specific gut pathogens without disrupting the broader microbial community; it remains largely in clinical development for gut applications but is advancing rapidly because of the antibiotic resistance crisis.
- Live biotherapeutic products represent the next generation of microbiome medicine, applying pharmaceutical-grade manufacturing and rigorous clinical trial standards to defined bacterial consortia;

most products are still years from commercial availability.

- Five questions separate evidence-based products from marketing noise: Was it tested in humans? In randomized trials? Replicated independently? Is the mechanism plausible? Is the evidence base free of funding from any dominant company? Products that pass all five for your specific concern are worth considering; products that fail most deserve skepticism.

9 The Microbial Frontier

In February 2004, a research vessel idled in the warm blue water of the Sargasso Sea, about 250 miles off the coast of Bermuda. A team led by geneticist Craig Venter lowered sampling equipment over the side, drew up 200 liters of seawater, and passed it through a fine filter. What remained on that filter was not just salt and microscopic debris. It was life, in numbers and variety that would astonish the scientific world when the results were published that same year.

The sequencing that followed did not target any single organism. Instead, Venter's team read every scrap of DNA caught in the filter, without knowing whose DNA it was or what organisms had produced it. The results rewrote the microbial census of the ocean. The filter held signatures from more than 1,800 microbial species, the vast majority previously unknown to science, along with more than 1.2 million novel genes. That single haul from a single sea nearly doubled the total number of known protein families in biology at the time. The ocean was not empty blue water. It was a living, incomprehensibly rich microbial ecosystem, and humans had been sailing across its surface for millennia without any idea what lived in the first centimeters of water beneath their hulls.

That expedition is often cited as the moment when metagenomics, the science of reading all the genetic material in an environmental sample at once, announced its ambitions to the world. Before the Sargasso Sea study, metagenomics had been a promising technique used in cautious, small-scale experiments. After it, the field became something else entirely: a wide-angle lens capable of surveying life at planetary scale. In the two decades since, that same technology has been applied to cancer patients, farm soil, Arctic permafrost, subway handrails, crime scenes, and livestock herds. This chapter is the story of where that lens has pointed, and what it has found.

Why the Frontier Matters

The earlier chapters of this book focused on your gut: the trillions of microbes living inside you, how they influence digestion, the immune system, mood, and disease risk. That story is compelling on its own. But metagenomics does not care about boundaries. The same techniques that reveal your gut microbiome composition can be applied to a handful of soil, a cup of seawater, a hospital surface, or a drop of blood. The technology is, at its core, a method for reading all the DNA in a sample without knowing in advance what you are looking for. That universality is what makes it so remarkable.

What is emerging from this broader application of metagenomics is a unified view of microbial life on Earth. The microbes in your gut are not isolated from the microbes in your backyard soil or those cycling carbon in the deep ocean. They are part of a single planetary system, connected through flows of genes, nutrients, water, and air. Understanding that system at a level of detail impossible before metagenomics is one of the defining scientific projects of this century. The practical stakes, for medicine, food security, the climate, and public safety, could hardly be higher.

Metagenomics and Cancer

The gut microbiome's relationship with the immune system is one of the most thoroughly documented findings in recent microbiome science. Roughly 70 percent of the body's immune cells reside in and around the gut, and the microbial communities there are in constant conversation with those cells, calibrating how aggressively or tolerantly the immune system responds to perceived threats. For decades, this relationship was studied mainly in the context of autoimmune conditions and infection. Then oncologists began noticing something unexpected: cancer patients with more diverse, healthier gut microbiomes were responding better to a class of treatments that amplifies the immune system's ability to attack tumors.

That class of treatments is called immune checkpoint inhibitors. Tumors often hijack molecular "brakes" in the immune system to protect themselves from immune attack. Checkpoint inhibitors block those brakes, freeing immune cells to recognize and destroy cancer. For some patients, the results are remarkable. For others, the same treatment does almost nothing, and for years oncologists struggled to explain why.

Research teams at the MD Anderson Cancer Center in Texas and the Gustave Roussy Institute in France independently found that patients with higher gut microbial diversity, particularly those with higher levels of certain species, including Akkermansia muciniphila and Bifidobacterium, had significantly better responses to checkpoint inhibitor therapy. Patients whose gut communities had been depleted by prior antibiotic use were far less likely to benefit. In some studies, patients with favorable microbiome profiles were more than twice as likely to show sustained tumor regression.

The mechanisms are still being worked out, but several pathways are well established. Certain bacteria produce short-chain fatty acids and other metabolites that directly modulate immune cell activity. In contrast, others stimulate the production of signaling molecules that prime immune cells for a more vigorous anti-tumor response.

Diagram 9.1. Cancer Immunotherapy and the Gut

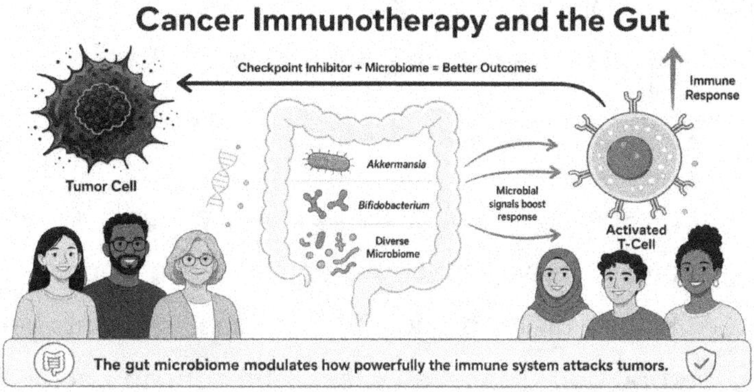

Several clinical trials are now testing whether correcting an unfavorable microbiome before treatment, through dietary interventions, probiotics, or fecal microbiota transplantation, can improve outcomes. Early results are encouraging, though the field is still learning which interventions work best for which patients and tumor types. The metagenomics dimension extends into what researchers call liquid biopsy metagenomics: detecting cancer-associated microbial DNA signatures in blood or stool before tumors become clinically visible. Certain bacteria, including Fusobacterium nucleatum, are present at elevated concentrations in colorectal tumors, and their DNA is detected in stool and blood samples from affected patients. Researchers are exploring whether metagenomic panels that detect these microbial signatures could serve as early-warning tests, identifying patients who need further screening before a tumor is large enough to cause symptoms or appear on imaging. This is not yet a clinical standard, but the research trajectory is clear.

The Soil Microbiome

Pick up a teaspoon of healthy agricultural soil, and you are holding more microbial species than there are people on Earth. A single gram of rich topsoil contains between 10,000 and 50,000 distinct species, representing bacteria, archaea, fungi, protozoa, and viruses in a web of interactions so complex that scientists are still working out the basic rules. The soil microbiome is the engine that makes terrestrial life possible. Without it, plants cannot access the nutrients they need, dead organic matter accumulates without decomposing, and the fundamental chemical cycles that sustain ecosystems break down.

The functions that soil microbes perform are staggering. Nitrogen-fixing bacteria take atmospheric nitrogen, which plants cannot use directly, and convert it into forms plant roots can absorb. Without this microbial process, terrestrial plant life and the food supply that depends on it would be severely constrained. Decomposer microbes break down dead organic matter, recycling carbon, phosphorus, and other essential nutrients back into circulation. Mycorrhizal fungi form intimate partnerships with plant roots, dramatically expanding the root system's effective surface area. Certain microbial communities also suppress plant pathogens, protecting crops from disease without chemical intervention.

Diagram 9.2. The Soil Microbiome at Work

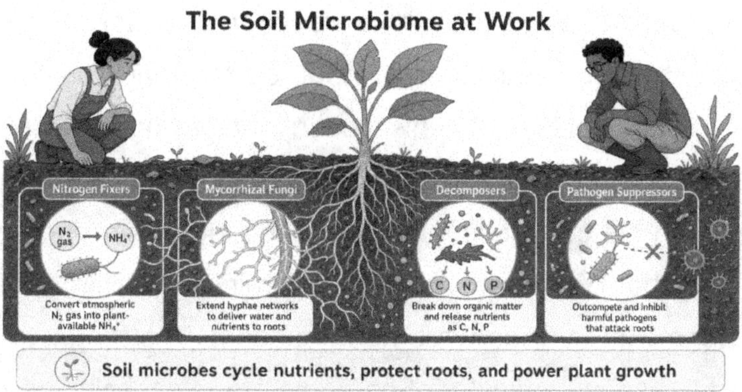

Industrial agriculture over the past century treated soil primarily as a physical medium for delivering synthetic fertilizers. Heavy tillage disrupted soil structure and the microbial communities within it. Synthetic nitrogen fertilizers reduced selection pressure for nitrogen-fixing bacteria. Broad-spectrum pesticides killed non-target microbial communities alongside pests. Monoculture cropping eliminated the diversity of root exudates and organic inputs that support diverse microbial life. The cumulative result has been a dramatic reduction in soil microbial diversity and function across large areas of agricultural land worldwide.

The regenerative agriculture movement, which encompasses reduced tillage, cover cropping, composting, and crop rotation, is explicitly aimed at restoring soil health and with it the soil microbiome. Consider a composite story drawn from many real farmers working with soil health programs: Darnell, a corn and soybean grower in central Illinois, began working with a regenerative agriculture consultant after watching his input costs climb for five consecutive years. When his consultant arranged a soil metagenomic test, the results were sobering. Microbial diversity in his conventionally managed fields was roughly one-third that of a nearby reference prairie site. Nitrogen-fixing bacterial families were nearly absent. The fungal-to-bacterial ratio, an indicator of carbon storage capacity, was low.

Over three years, Darnell adopted cover cropping, eliminated fall tillage, and began composting a portion of his corn stover. Follow-up metagenomic tests showed measurable improvements: nitrogen-fixing populations had recovered, fungal diversity had increased, and soil organic matter had risen by half a percentage point. His fertilizer costs dropped by about 15 percent by the third year because the microbial community was doing more of the nutrient cycling. The metagenomics testing served as both a diagnostic tool and a feedback mechanism, making improvements in soil biology visible long before the bottom line confirmed them.

Ocean Metagenomics

Venter's 2004 Sargasso Sea expedition opened a door that has not been closed since. The Tara Oceans project, which launched in 2009 and spent four years sailing around the world, collected metagenomic samples from hundreds of stations across all major ocean basins. The resulting dataset has produced discoveries that continue to emerge years after the last sample was collected.

The numbers are almost incomprehensible. The world's oceans harbor an estimated 10^29 microbial cells. Marine microbes, primarily bacteria, archaea, and single-celled algae called phytoplankton, produce approximately half of the oxygen in Earth's atmosphere. A microorganism in the ocean produces every second breath you take. These same marine microbes are primary drivers of the ocean carbon cycle, fixing carbon dioxide through photosynthesis, cycling it through the food web, and exporting a portion of it to the deep ocean where it can remain stored for centuries. This biological carbon pump is one of the most important mechanisms regulating the planet's climate.

Diagram 9.3. The Ocean's Microbial Engine

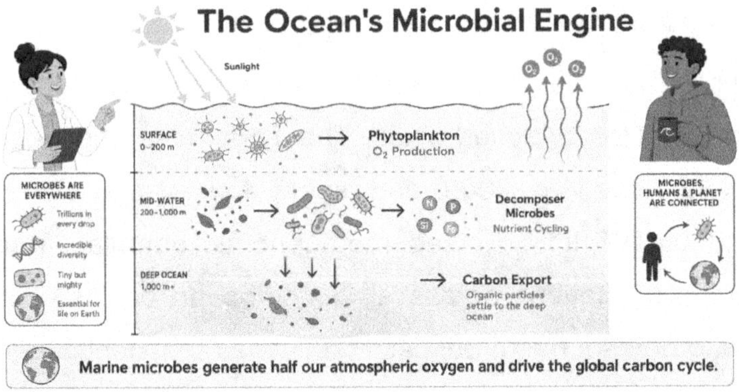

Metagenomics has revealed that ocean microbial communities vary enormously by depth, temperature, nutrient availability, and light exposure, forming distinct ecological zones that function differently in the carbon cycle. Surface waters in nutrient-rich upwelling zones support dense, diverse photosynthetic communities. The deep ocean harbors different communities that derive energy from chemical reactions rather than sunlight. Climate change is already altering these communities in measurable ways: ocean surface temperatures have risen, warm water holds less dissolved oxygen, and in some regions warming has reduced mixing between surface and deep water layers, cutting off nutrient supply to surface communities.

What metagenomics adds to this picture is precision. Before metagenomic surveys, ocean ecologists tracked microbial changes mainly through proxy measurements such as chlorophyll concentrations and oxygen levels.

Metagenomics allows researchers to track specific strains, gene pathways, and metabolic activities at a global scale. They can now detect, in real time, whether nitrogen fixation genes are becoming less abundant in a given ocean region, or whether microbial pathways that produce dimethylsulfide, a gas that influences cloud formation, are being disrupted by warming. Building ocean metagenomics monitoring programs to serve as early warning systems is now a priority for several major international research initiatives.

Metagenomics and Climate Change

The connection between microbes and climate runs deeper than ocean photosynthesis. Microbes are the planet's primary decomposers, responsible for breaking down dead organic matter and returning its components to circulation. When they decompose organic matter, they produce gases including carbon dioxide and methane that enter the atmosphere. Microbial decomposition is therefore a fundamental biological control on greenhouse gas concentrations, not merely a contributing factor.

Nowhere is this more vividly illustrated than in Arctic and subarctic permafrost zones. Permafrost is ground frozen for at least two consecutive years, and vast expanses of Arctic ground have been frozen for thousands of years. Within that frozen ground lies an enormous quantity of organic carbon, the accumulated remains of past plants and animals, preserved by the cold. Estimates suggest Arctic permafrost soils contain roughly twice as much carbon as currently exists in Earth's atmosphere. The Arctic is warming at roughly two to four times the global average rate, and permafrost is thawing.

Diagram 9.4. The Permafrost Feedback Loop

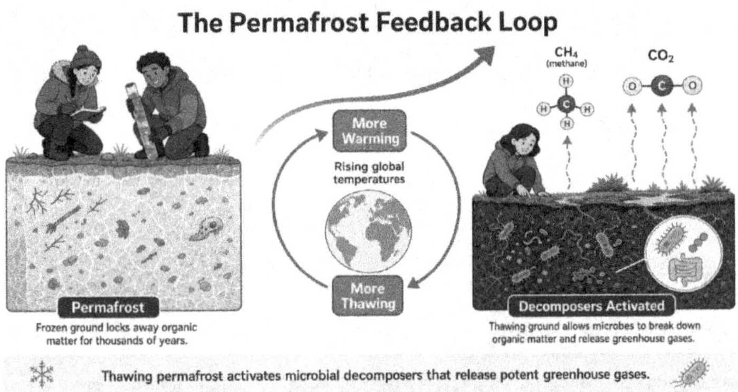

When permafrost thaws, microbial communities that have been metabolically dormant at freezing temperatures awaken and begin consuming the organic carbon previously locked in the ice, producing carbon dioxide and methane as byproducts. Methane is particularly concerning because its warming potential over a 20-year timeframe is roughly 80 times that of carbon dioxide. The worry is not just that thawing permafrost will release more greenhouse gases, but that it will do so in a self-amplifying feedback loop: warming causes thawing, which activates microbial decomposition, which produces more greenhouse gases, which causes more warming.

Metagenomics allows scientists to monitor this process with mechanistic detail. By sequencing microbial communities in permafrost soils at different stages of thaw, researchers can identify which organisms produce methane (primarily archaea called methanogens), which produce carbon dioxide, and which unexpectedly consume methane, partially offsetting its release. This matters enormously for building accurate climate models. The answer is not simply "more warming equals more carbon release"; the specific microbial community, temperature, moisture, and the composition of organic matter all shape the outcome.

Microbial methane production is not confined to the Arctic. Livestock, particularly ruminants such as cattle and sheep, harbor methanogens in their digestive systems that produce methane as a byproduct of fermenting plant material. Global livestock agriculture accounts for approximately 14.5 percent of anthropogenic greenhouse gas emissions, with enteric fermentation contributing the largest share. Metagenomics has characterized the gut microbiomes of cattle, identifying which methanogenic archaea are most responsible and revealing variation between individual animals, opening research pathways toward low-emission livestock genetics, methane-suppressing feed additives, and rumen microbiome modification.

Forensic Metagenomics

Every person who moves through the world leaves behind a microbial trail. The skin sheds cells daily, and each carries bacteria from the unique community that colonizes that individual's skin. Researchers at the University of Colorado Boulder showed this in a direct demonstration: by sequencing a keyboard's microbial community, they could identify its regular user from a pool of candidates with high accuracy. This is forensic metagenomics — the application of environmental DNA sequencing to questions of identity, time, and place.

Diagram 9.5. Forensic Microbial Signatures

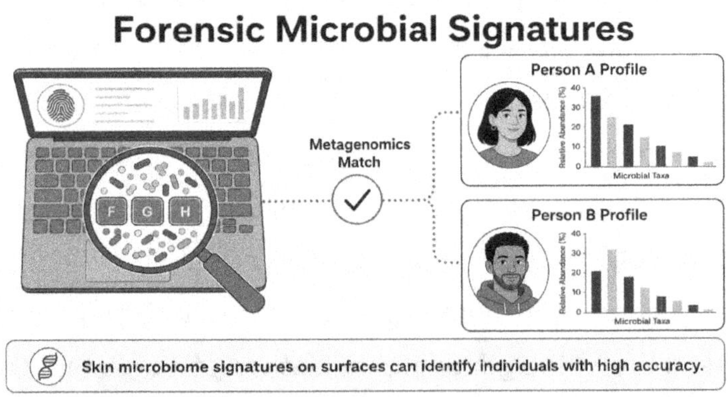

The applications extend beyond identification. Crime scene investigation gains a complement to human DNA: microbial succession on a body after death follows a predictable timeline that can help estimate time of death, paralleling the use of insect succession in forensic entomology. Food safety investigation benefits similarly — metagenomics can characterize the entire microbial signature of a contaminated product, allowing investigators to match it to specific production facilities. The technique is already in use in high-profile outbreak investigations.

The Built Environment Microbiome

Most people spend roughly 90 percent of their waking hours indoors. The buildings we inhabit are ecosystems. The MetaSUB consortium — a global network coordinated out of Weill Cornell Medicine — sampled subway systems, airports, and public spaces in more than 60 cities worldwide and found that each city has a characteristic microbial signature: stable, reproducible, and as geographically specific as its architecture. The Tokyo subway's microbiome is measurably different from New York City's, which is measurably different from the London Underground.

Diagram 9.6. The MetaSUB City Map

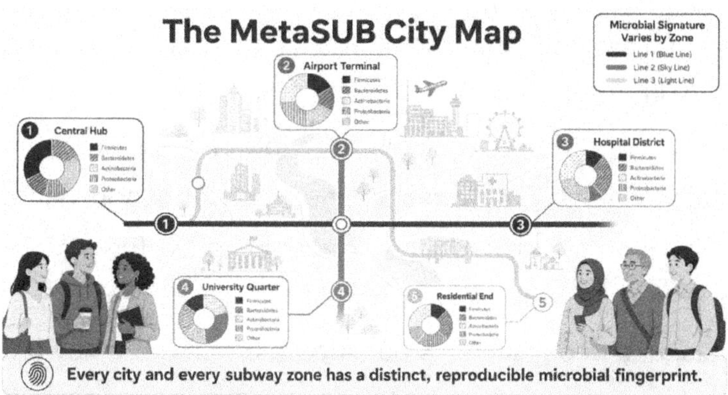

The health implications are real but not alarming. Buildings with greater microbial diversity — achieved through natural ventilation, occupant contact with outdoor environments, and diverse human populations — are associated with lower rates of sick-building syndrome and healthier indoor microbiomes among occupants. Window design and ventilation systems shape microbial communities in ways that can matter for respiratory health. The design of built environments is, among other things, a public health intervention.

One Health: The Planet as a Single Microbial System

The sections above have visited the microbiomes of cancer patients, farm soils, ocean water, thawing permafrost, subway handrails, and hospital wards. On the surface, these seem like separate topics connected only by the shared technology of metagenomics. But there is a deeper organizing principle that ties them together: the One Health framework.

One Health is the recognition, now formally endorsed by the World Health Organization, the Food and Agriculture Organization of the United Nations, and the World Organization for Animal Health, that the health of humans, animals, and ecosystems is not separate matters. They are dimensions of a single integrated system, and what happens in one dimension reverberates in the others. The COVID-19 pandemic provided the most vivid recent illustration: a virus originating in an animal reservoir moved into human populations through zoonotic spillover and within months disrupted economies and health systems worldwide. But the connections run far deeper and operate on far longer timescales than pandemic events.

Diagram 9.7. One Health Triangle

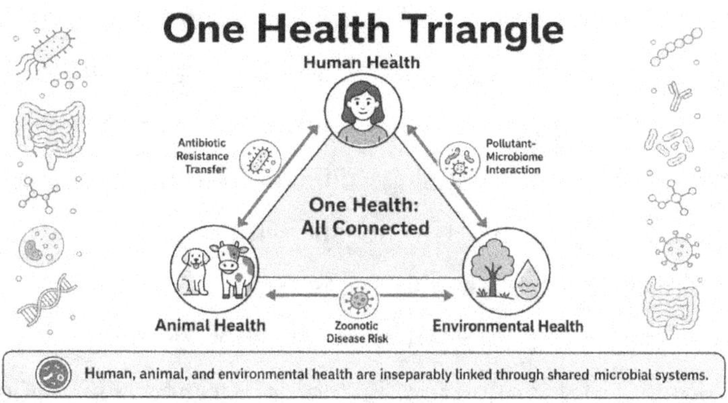

Antibiotic resistance is one of the clearest examples of One Health dynamics at play in microbial systems. Antibiotics are used not only in human medicine but in enormous quantities in animal agriculture, both therapeutically and, in many countries, to promote growth. Antibiotic-resistant bacteria that evolve in animal gut microbiomes can be transferred to humans directly through food handling, or indirectly through contaminated water and soil. Resistance genes can move between entirely different bacterial species via horizontal gene transfer, the sharing of genetic material that makes resistance a community-level property rather than a species-level one.

Metagenomics has made it possible to track antibiotic resistance genes, collectively called the resistome, across the One Health continuum. Studies have traced resistance gene signatures from livestock manure into the soil microbiome, from soil into agricultural runoff, from runoff into streams and rivers, and from waterways into the gut microbiomes of people who live downstream. This is not a theoretical concern: resistance genes first identified in livestock pathogens have been found in human clinical isolates of bacteria causing difficult-to-treat infections. Metagenomics provides a map showing exactly how this movement occurs and at what rates.

Zoonotic disease surveillance is another domain where One Health metagenomics is demonstrating its value. Identifying which animal populations harbor pathogens with pandemic potential before they make the jump to humans requires surveillance at scale, something economically infeasible before metagenomics reduced the cost and complexity of environmental DNA analysis. Global initiatives are now deploying metagenomics-based surveillance in wildlife populations and livestock operations in high-risk regions, aiming to detect novel pathogen signatures before they establish themselves in human populations.

The One Health vision is not just descriptive but prescriptive. If human health, animal health, and environmental health are inseparably linked, then managing them in separate silos will always produce solutions in one domain that create problems in another. Antibiotics that save human lives create resistance that undermines them. Pesticides that protect crops damage the soil microbial communities that crops depend on. The One Health framework demands integrated governance and science, and metagenomics is the common language that enables integration.

Diagram 9.8. The Ripple of Metagenomics

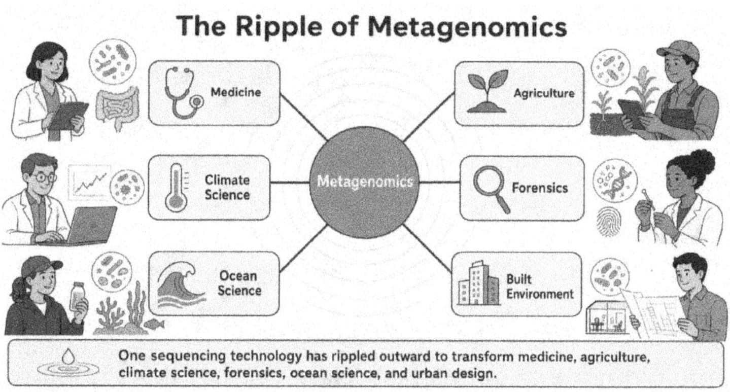

The most hopeful vision of One Health metagenomics is the design of what could go right. Farmland managed to support rich soil microbial communities might simultaneously improve crop yields, reduce synthetic input costs, sequester carbon, and reduce the burden of resistant pathogens in food animals. Urban environments designed with microbial diversity in mind might reduce allergies while lowering hospital-acquired infection rates. Ocean monitoring programs tracking microbial health might provide early warning intelligence needed to protect fisheries and assess climate mitigation at the level of biological reality.

These are aspirational goals, and science always moves more slowly than aspirations suggest. But the technology that makes them imaginable, the ability to read all the DNA in any environment at increasing speed and declining cost, is already in hand. The question is not whether metagenomics can transform our understanding and management of the microbial systems that support life on Earth. It already has. The question is how quickly and equitably the insights will be translated into practice.

Takeaways

The metagenomics revolution that began with Venter's Sargasso Sea expedition has expanded in ways its pioneers did not fully anticipate. Here is what this chapter's sweep across disciplines leaves you with:

- The gut microbiome is a meaningful predictor of how well cancer patients respond to immunotherapy. Clinical research is actively testing whether correcting the microbiome before treatment can improve outcomes, and metagenomic liquid biopsy approaches may eventually detect cancer-associated microbial signatures in blood or stool before tumors are clinically visible.

- Soil is alive in ways that industrial agriculture has largely ignored. A single teaspoon of healthy soil harbors more microbial species than there are people on Earth. Regenerative practices that restore soil microbial diversity are measurably improving soil function, reducing input costs, and beginning to contribute to carbon sequestration.
- The ocean's microbial communities produce approximately half of Earth's atmospheric oxygen and are central to the global carbon cycle. Climate change is already altering these communities, and ocean metagenomic monitoring programs are being built to provide early warning of disruptions to the ocean's biological engine.
- Microbial communities in permafrost soils, wetlands, and livestock digestive systems are significant drivers of greenhouse gas production. Metagenomics provides the mechanistic detail needed to build accurate models and develop mitigation strategies.
- Every person leaves a unique microbial signature on the surfaces they touch. Forensic metagenomics is developing tools to use those signatures for crime scene investigation, food safety tracing, and biosecurity

monitoring, alongside ethical and legal frameworks that must develop in parallel with the science.

- The built environments we inhabit have distinct microbial identities that influence our health. Designing buildings with microbial diversity in mind is an emerging frontier in architecture and public health.
- The One Health framework establishes that human health, animal health, and environmental health form a single integrated system. Metagenomics is the common language that enables tracking and management of this integrated system, from following antibiotic resistance genes through watersheds to monitoring wildlife populations for emerging zoonotic pathogens.

The microbial frontier is not a distant horizon. It is the soil under your feet, the air moving through your building, the water cycling through your watershed, and the communities living inside you. Metagenomics has given science, for the first time, the tools to read all of it.

10 Becoming Microbiome-Literate

Picture yourself in a doctor's waiting room in the year 2030. You dropped a small sealed tube into a collection kiosk near the front desk two days ago, and by the time you sit down across from your physician, a rapid metagenomic sequencing report is loaded on the shared screen between you. The report flags a drop in your Faecalibacterium prausnitzii levels, a bacterium that produces short-chain fatty acids that help calm intestinal inflammation. It also notes an overgrowth of a species the literature has associated with elevated colorectal cancer risk in people with your dietary pattern. None of this arrived with alarm bells. It arrived the way a cholesterol panel arrives: as data, contextualized, compared to your previous reading from eighteen months ago, and filtered through your physician's clinical judgment.

Your doctor leans forward and asks three questions. Did you take antibiotics recently? How has your vegetable intake been? Are you still getting outside most days? Together, you sketch a small set of changes: more diverse plant foods, a fermented food added to your weekly rotation, a prescription-grade microbial preparation that just cleared its third clinical trial. Your doctor notes that the elevated-risk species finding is early and nonspecific, not a diagnosis, and that the two of you will recheck in a year. You leave more informed than anxious, more empowered than overwhelmed.

Every piece of that scenario exists in some form today. Rapid metagenomic sequencing is real. Live biotherapeutic products are moving through regulatory pipelines right now. What is missing is the integration: validated clinical interpretation, reimbursement frameworks, equitable distribution, and physician training. This final chapter is about the distance between where microbiome science stands and where it is heading, and what you, as a now-informed reader, can do with that knowledge in the meantime.

Why This Chapter Exists

You have spent nine chapters building a foundation: what the microbiome is, how it forms in the earliest hours of life, how diet shapes it across a lifetime, how it communicates with your brain, how it trains your immune system, and how disruptions to it connect to conditions ranging from irritable bowel syndrome to metabolic disease to depression. You have examined the evidence for probiotics and fermented foods, learned what FMT is and when it works, and explored the frontier of personalized microbiome medicine. That foundation is not just intellectual background. It is the prerequisite for everything in this chapter.

Becoming microbiome-literate is not a state you arrive at and stay in. It is an active practice combining genuine curiosity with disciplined skepticism. It means knowing when a headline deserves your attention and when it overreaches the evidence, what questions to bring to your physician, and how to read a microbiome test report without dismissing it or overreacting. And it means understanding that you are living at an extraordinary moment in medicine, where the tools to make sense of the trillions of organisms inside you are being built in real time.

Consider the composite story of Richard, a recently retired high school science teacher in his early sixties who has not had a full physical exam in four years. He feels generally well and walks his dog every morning, but he has noticed bloating after meals over the past year and read an article suggesting his gut bacteria might be to blame. He printed the article out and is now sitting in the waiting room, wondering whether to hand it to his doctor or leave it in his pocket. He wants to ask about gut health but is not sure how to start the conversation or what tests are worth requesting. Richard is prepared, curious, and just a little uncertain. This chapter is for him, and for you.

What the Evidence Actually Supports

Not all microbiome advice is created equal. The field uses a rough evidence hierarchy that ranks claims from the most robustly supported to the weakest. Understanding where a recommendation sits does not tell you whether to follow it, but it tells you how much weight to put on it and how much further evidence is needed.

At the strongest tier sit interventions supported by multiple large, well-designed human trials that replicate across different populations and labs. These are the same lifestyle foundations that cardiologists, endocrinologists, and gastroenterologists have recommended for decades, now with a microbiome explanation layered underneath. Eating at least 30 different plant foods per week, maximizing dietary fiber, limiting ultra-processed foods, exercising most days, getting 7 to 9 hours of sleep, spending time outdoors, and avoiding unnecessary antibiotics each has strong, converging human research support. They are not exciting new interventions. They are the most defensible foundation you can build.

Diagram 10.1. The Three Evidence Tiers

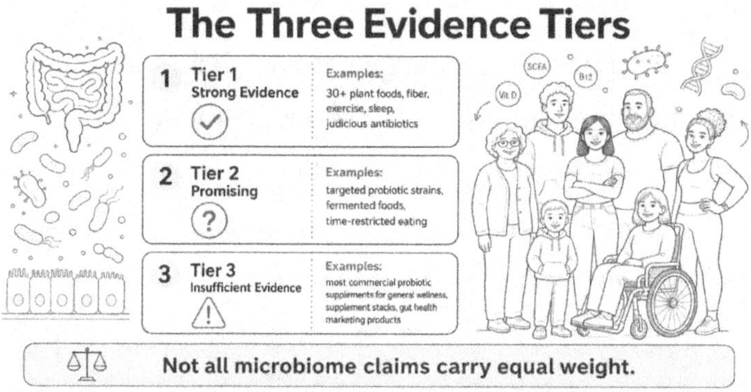

At the second tier sit interventions with genuine promise but where the human evidence is limited in scale, inconsistent across trials, or dependent on individual variables science has not yet fully characterized. Targeted probiotic strains for specific clinical indications, such as Lactobacillus rhamnosus GG for antibiotic-associated diarrhea or Bifidobacterium infantis for infant colic, sit here because the evidence is real but context-dependent and does not generalize to all patients or products. Regular consumption of fermented foods, including yogurt, kefir, kimchi, and sauerkraut, also belongs in this tier: a well-designed trial showed measurable increases in microbiome diversity and reductions in immune markers in healthy adults following a high-fermented-food diet, but the effect likely varies by baseline diet and individual microbiome composition.

Time-restricted eating is another second-tier candidate. Restricting eating to a consistent daily window can favorably shift microbiome composition and improve metabolic markers in shorter studies. Still, the optimal window, the population that benefits most, and the durability of the effect remain unsettled. Pursue second-tier interventions with reasonable expectations and without substituting them for first-tier foundations.

At the third tier sit most commercial gut health products: probiotic blends marketed for general wellness, prebiotic powders with minimal clinical backing, and supplement stacks promising to "balance your microbiome" without a defined outcome measure. The way to evaluate any third-tier product is direct: what randomized controlled trial in humans supports this product's specific claim? If the answer is "none," that is your answer.

How to Read a Microbiome Test Result

Tens of thousands of people now receive consumer microbiome test reports each year without any preparation for interpreting what they actually mean. The combination of official-looking numbers and unfamiliar species names can tilt someone toward either dismissal or overreaction. What you need is a practical reading framework.

The first thing to understand is what a consumer test is actually measuring. Most use 16S rRNA gene sequencing, which identifies bacterial groups by targeting a specific genetic marker. This method is cost-effective but, in many cases, provides genus- or family-level identification rather than species-level precision, and it cannot detect viruses, fungi, or all bacterial strains equally. Some newer tests use shotgun metagenomic sequencing, which provides more resolution but is more expensive and computationally intensive. Neither method tells you what those organisms are doing in your gut; they tell you who is present and at roughly what relative abundance.

Diagram 10.2. How to Read a Test Report

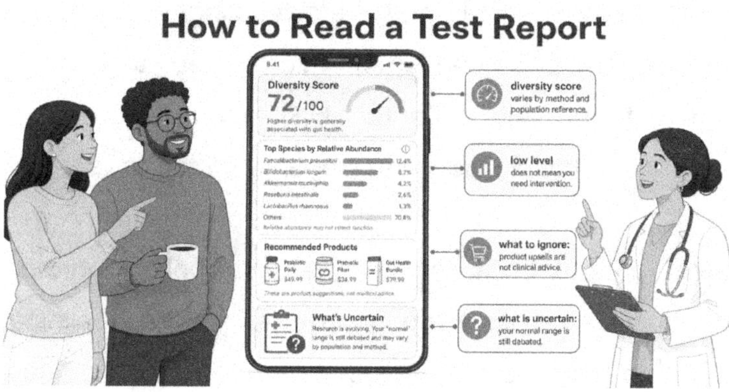

The second thing to understand is what a diversity score does and does not tell you. Higher alpha diversity is broadly associated with better metabolic and immune health at the population level, but this association is statistical and probabilistic. A person with moderate diversity might be in excellent health, and a person with a very high diversity score might have an active inflammatory condition. Your score is a data point, not a verdict.

The third thing to examine critically is how the test handles individual species findings. If a report says your level of a beneficial species is "low," ask: low compared to what reference population? Is that population demographically matched to you? Is there a validated clinical intervention for raising that species, or is the recommended action to buy the company's probiotic? Consumer test companies have a financial incentive to generate actionable-seeming recommendations, and the most common action they direct you toward is a purchase. Verify any recommendation independently before acting on it.

Richard received a consumer microbiome test as a birthday gift from his daughter. His report flagged "low Akkermansia muciniphila" and recommended a specific probiotic blend. He found that while low levels of this bacterium have been associated with metabolic syndrome in population studies, there is no well-validated clinical threshold for what "low" means in a healthy sixty-two-year-old without metabolic disease, and that the probiotic his test recommended contained no Akkermansia at all. He brought this to his appointment not as a confrontation but as a question: what should I make of this?

A final red flag is the language a test company uses. Phrases like "your microbiome is imbalanced" or "we detected a dysbiosis" are significant warning signs. As described in Chapter 2, dysbiosis is not a precisely defined clinical entity with agreed diagnostic criteria. A company using that language without clinical specificity is overstating what a single sequencing snapshot can tell you.

Conversations to Have With Your Doctor

Microbiome science has advanced faster than continuing medical education keeps physicians current. A primary care doctor finishing medical school in 2010 received essentially no training in the gut-brain axis, live biotherapeutics, or metagenomics-guided clinical decision-making. The practical implication is that you may know more about recent microbiome research than your doctor does, and framing your questions as collaborative inquiry rather than challenge will produce better outcomes for both of you.

The first conversation worth having is about antibiotics. Antibiotic stewardship remains an active challenge in outpatient medicine. The microbiome evidence adds a powerful additional argument for restraint: a single course of broad-spectrum antibiotics can measurably reduce microbiome diversity, and in some individuals, the effects persist for months. The questions worth asking are simple. Is this infection bacterial or viral? If bacterial, what is the narrowest-spectrum antibiotic appropriate for the likely pathogen? Is watchful waiting for 48 to 72 hours a reasonable option for a mild infection in an otherwise healthy adult?

Diagram 10.3. Five Questions for Your Doctor

The second conversation is about your specific condition. If you are managing IBS, IBD, type 2 diabetes, a mood disorder, an autoimmune condition, or recurrent infections, ask whether your physician is tracking microbiome-relevant research for that area and whether any microbiome-informed approaches are relevant to your care. Questions like "Are there dietary changes with evidence for my specific condition?" or "Are there probiotic strains with good trial data for IBS-D?" are entirely reasonable.

The third conversation is about testing and staying current. Clinical-grade microbiome tests are becoming available in some specialty settings, particularly gastroenterology and oncology. A consumer test is not a clinical test, but the conversation signals to your doctor that you are engaged. You might ask: "Is there a source you use to stay current on microbiome research?" Physicians who are good care partners generally welcome engaged patients.

The fourth conversation is about your daily habits. Your doctor can prescribe medication, but the thirty-plus plant foods per week, sleep hygiene, outdoor time, and stress management are lifestyle changes where you are the agent. A good physician will want to know whether you are implementing them and what barriers you face. Richard ended up having a forty-minute conversation covering his diet, his sleep since retirement, his stress around a family matter, and a potential referral to a dietitian specializing in gut health. None of that required new technology. It required only that he came prepared to engage.

The Future of Microbiome Medicine

The next decade of microbiome medicine will be shaped by at least three converging developments, each already visible in the current research landscape. Understanding where they stand will help you calibrate your expectations, recognize genuine advances, and distinguish them from hype.

The first development is the maturation of live biotherapeutic products, or LBPs. These are FDA-regulated drugs consisting of living microorganisms, developed through the same rigorous clinical trial process as any pharmaceutical. A probiotic supplement does not require proof of efficacy to be sold. An LBP has completed a defined clinical trial program, proven its efficacy for a specific indication in a randomized controlled trial, and received regulatory approval before it can be prescribed. Vowst, the first oral LBP approved by the FDA, is a microbiome-restoration product for recurrent Clostridioides difficile infection and marks the beginning of a new regulatory category with enormous potential reach.

Diagram 10.4. The 2030 Microbiome Clinic

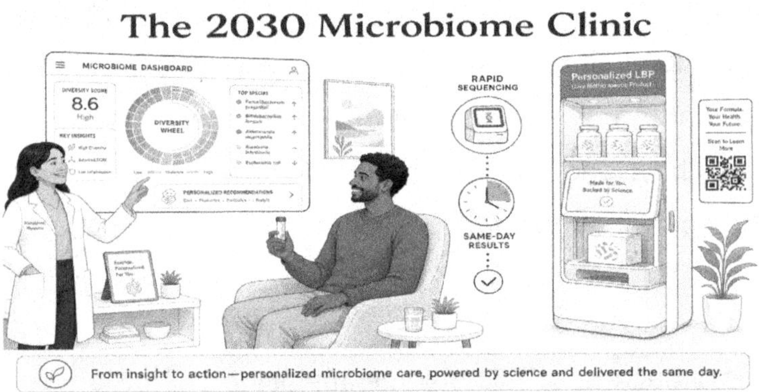

The second development is the integration of microbiome profiling into cancer care. Multiple large observational studies and several clinical trials now suggest that a patient's microbiome composition at the start of immunotherapy is a meaningful predictor of treatment response. Patients with high microbiome diversity and adequate levels of Faecalibacterium prausnitzii and several Ruminococcus species appear more likely to respond to checkpoint inhibitor therapies for melanoma, lung cancer, and bladder cancer. Researchers are testing whether modifying the microbiome before immunotherapy can improve outcomes in non-responders. Early results in small trials are intriguing, and larger trials are underway.

The third development is the integration of microbiome monitoring into routine preventive care. The vision is a standardized microbiome panel, validated for clinical use, reimbursed by insurance, and administered alongside routine bloodwork. Such a panel would not diagnose specific diseases; it would track meaningful trends over time and flag early signals that warrant follow-up. Several academic medical centers are already piloting longitudinal microbiome monitoring in research cohorts. The clinical translation will require time and rigorous validation, but the direction of travel is clear.

Beyond these three, the microbiome field is generating promising early research in Alzheimer's disease prevention, gestational health, pediatric autoimmune conditions, and aging. The animal models are compelling, the biological mechanisms are increasingly understood, and the first human trials are producing findings that justify continued investment. What these areas lack, for now, is the weight of replicated large-scale human evidence needed for clinical implementation. That evidence is coming.

Equity and Access: Who Gets Microbiome Medicine?

Medical breakthroughs and the populations who most need them have a complicated history. Dialysis, targeted cancer therapy, and HIV antiretrovirals all followed a common early trajectory: they arrived first for patients with resources, and broader access came slowly, imperfectly, and only under sustained advocacy pressure. Microbiome medicine is at an early enough stage that the equity questions are still open. How those questions are answered will determine whether the microbiome revolution becomes a genuine public health advance or a premium service available only to a relatively small, already advantaged population.

The challenge begins with clinical trial populations. As this book described in Chapter 3, most of the microbiome reference data on which "normal" diversity benchmarks are built has come from white, Western, university-affiliated populations. The same demographic skew applies to many of the clinical trials that generated the evidence base for probiotics, FMT, and LBPs. When a diverse patient is told her microbiome diversity score is "below average," the reference range used may not adequately represent people with her ancestry, diet, or geographic background. Correcting this requires deliberate effort: funding trials in diverse populations, incentivizing enrollment from historically excluded communities, and building reference databases that reflect the full range of human microbiome variation. None of this happens automatically; it requires advocacy and funding priorities that researchers, regulators, and patients need to push for together.

Diagram 10.5. Equity Gap

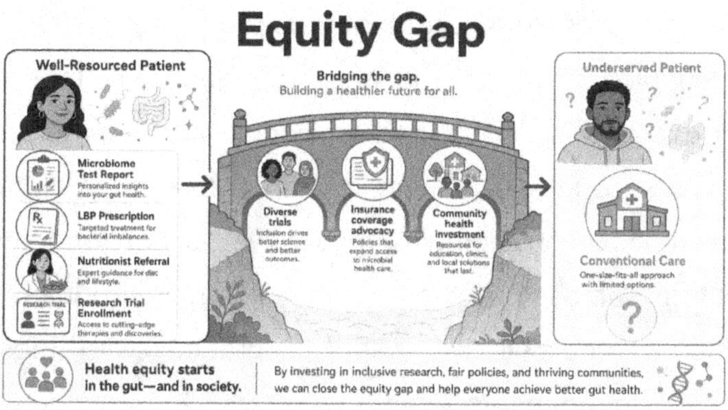

Insurance coverage is the second structural barrier. Live biotherapeutic products, even those with FDA approval, are not automatically covered by insurance. Vowst entered the market at a list price that placed it out of reach for uninsured patients and created reimbursement uncertainty for many with standard plans. As the LBP pipeline expands, the question of who pays will be central. Consumer microbiome testing is rarely covered, and clinical-grade testing, when it arrives, will face the same coverage battles that genetic testing and advanced imaging have already fought. The people least burdened by the diseases these technologies might help are typically the most likely to access them early. At the same time, those carrying the highest burden of inflammatory and metabolic disease wait longest.

The dietary recommendations at the top of the evidence hierarchy are theoretically available to everyone, but the conditions that make them possible are not equally distributed. Access to affordable fresh produce is limited in many low-income urban and rural communities. Time for regular exercise is constrained for people working multiple jobs. Chronic stress, which disrupts the microbiome through multiple physiological pathways, is not randomly distributed across society. A genuinely equitable microbiome medicine will need to grapple with these structural determinants, not just with the molecular interventions that arise from studying them.

What can you do about equity in microbiome medicine? More than you might expect. Supporting citizen science projects that prioritize diverse sampling contributes to better reference data. Asking your insurer explicitly about coverage for emerging microbiome therapies creates a paper trail that matters in coverage decisions. Contacting your elected representatives about funding for microbiome research in underserved populations shapes the advocacy environment in which research priorities are set. Each of these actions is modest on its own. Together they add up.

Citizen Science: You Can Participate

One of the genuinely remarkable features of the microbiome research landscape is the degree to which ordinary people have contributed to it directly. The American Gut Project, which began at the University of California, San Diego, in 2012, enrolled more than 15,000 participants who contributed stool samples and completed detailed health and dietary questionnaires. The findings from that dataset have supported dozens of published studies, refined the understanding of diet-microbiome relationships, and helped identify microbiome signatures associated with various health conditions.

Harold, the retired teacher from the previous chapter, has reached this point in the book and has made two practical decisions. First, he has bookmarked the American Gut Project page; he likes the idea that the next sample he might mail in would help build the reference dataset that future Harolds will benefit from. Second, he has written a single question on the back of an index card for his next annual physical: 'When you order my labs, is there anything microbiome-adjacent you would now look at differently than you would have ten years ago?' He is not expecting a long answer. He is testing whether the literacy he has built in these pages can travel into a real ten-minute conversation. That is exactly the kind of test this chapter is asking him to run.

The Earth Microbiome Project maps microbial life across ecosystems globally, from soil and seawater to plants, animals, and human bodies, enlisting contributors from more than forty countries. MetaSUB, the Metagenomics and Metadesign of Subways and Urban Biomes project, coordinates sample collection from public transit systems in dozens of cities worldwide, building a picture of the microbial life that urban environments share. ZOE, the largest personalized nutrition study in history, has enrolled hundreds of thousands of participants and is generating longitudinal data on how individual microbiome composition, genetics, and metabolic responses to food interact.

Diagram 10.6. Citizen Science Projects

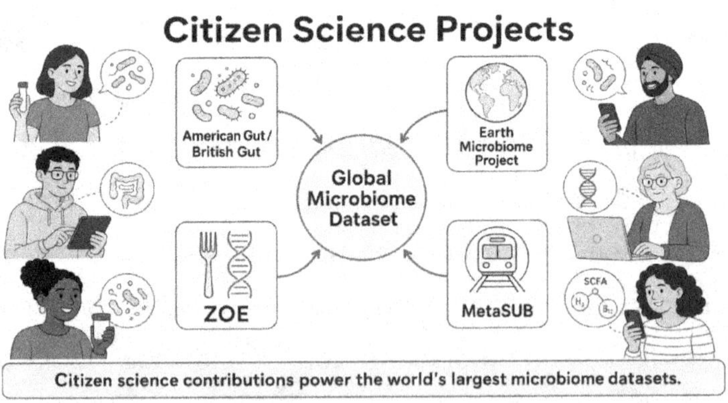

Participating in citizen science is straightforward. Most projects provide a sample kit, collection instructions, and a questionnaire covering diet, medications, sleep, stress, and health conditions. You return the sample, your data is incorporated into the dataset, and in some cases you receive a report of your own results. Each additional participant from an underrepresented demographic, a rural area, a non-Western diet pattern, or an unusual health background adds information that no amount of increased sequencing depth on already-enrolled participants can replace.

Good entry points include the ZOE website, academic medical centers conducting gut health studies in your region, and university-affiliated citizen science platforms that list active enrollment opportunities. You do not need a science background or any equipment beyond what the study provides. One caution: read the consent form carefully. Understand how your data will be stored, who will have access to it, whether it will be de-identified, and whether you retain the right to request deletion. Reputable projects affiliated with academic institutions have robust ethics review and clear data governance policies.

11 The Microbiome in Daily Life

It is 6:40 on a Thursday morning, and Elena is standing in her kitchen deciding whether the day ahead will be another exercise in improvisation or the beginning of a more stable routine. She slept six hours because she stayed up finishing work. Breakfast could be oatmeal with berries, or coffee and nothing else. Lunch will depend on whether meetings run long. By evening, she will either take a walk, collapse on the couch, or open a delivery app and order whatever sounds easiest. None of these decisions feels dramatic on its own, yet together they form the texture of the microbial world inside her. This is the level at which the microbiome becomes most relevant to ordinary life: not as a miracle hidden in a capsule, but as a steady response to the patterns repeated across days and weeks.

What many readers want by this point in the book is not another abstract claim that the microbiome matters. They want to know what to do on a normal Wednesday, after a stressful week, during travel, after a stomach bug, or when motivation is low. This chapter answers that question carefully. It does not claim that every daily choice rewrites your microbiome in a clinically meaningful way. It argues something both more modest and more useful: daily routines create the conditions in which a resilient microbial ecosystem is more or less likely to thrive, and those routines are among the few things most people can actually influence.

The Everyday Ecology of Health

One reason microbiome communication goes wrong is that it often treats the gut as if it responds to isolated superfoods or villain ingredients rather than to patterns. In reality, your microbiome experiences your life as a sequence. It registers whether fiber arrives repeatedly or rarely, whether sleep is regular or chaotic, whether movement is ordinary and frequent or absent for days at a time, whether stress is punctuated by recovery or allowed to become a permanent background condition. The question is not whether one salad "fixes" the gut or one bad weekend "damages" it. The more accurate question is what kind of environment your habits are creating over time.

An ecological way of thinking helps here. A healthy garden is not produced by one perfect watering but by repeated conditions that allow many organisms to coexist and recover after disturbance. The same principle applies to the gut. Diversity in food sources, especially plant foods with varied fibers and polyphenols, tends to create a richer substrate for microbial metabolism. Regular sleep and circadian rhythm stabilize hormone signaling, appetite timing, and bowel regularity. Routine movement supports motility and metabolic health. None of this is glamorous, which is one reason it is often displaced by more marketable advice. But in practice, these are the levers that matter most for most readers.

Diagram 11.1. A Day in Microbiome Time

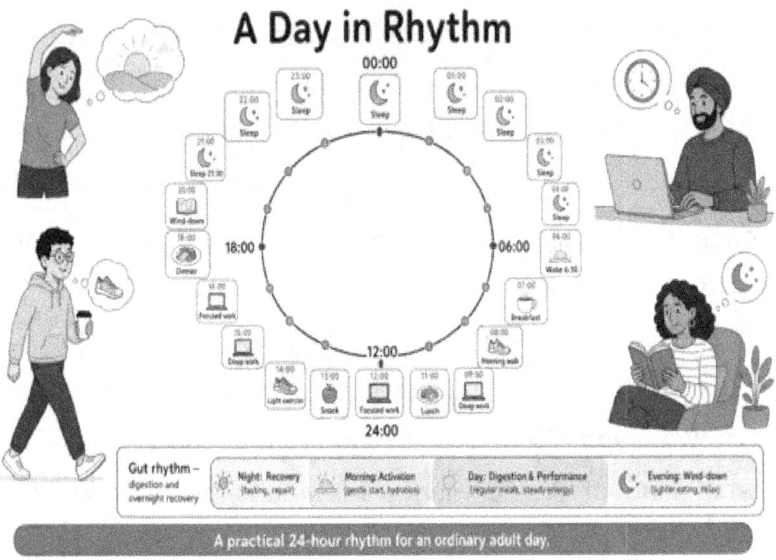

Meals as Repeated Signals

The gut does not interpret breakfast, lunch, and dinner as social categories. It interprets them as waves of substrate, chemicals, timing cues, and metabolic work. Meals rich in varied plant foods tend to provide the fermentable material from which gut microbes produce short-chain fatty acids such as butyrate, acetate, and propionate. Those metabolites are not abstract biochemical trivia; they help nourish colon cells, modulate inflammation, and participate in the broader signaling network linking the gut to immunity and metabolism. This is why the question of diet in microbiome health is less about identifying one miracle ingredient than about building enough repetition around known supportive patterns that your internal ecosystem can rely on them.

A useful practical shift is to think in terms of weekly diversity rather than daily perfection. A person who eats oats, lentils, berries, greens, beans, seeds, yogurt, herbs, and a wide range of vegetables over the course of a week is giving the microbiome a much richer menu than someone whose diet is technically healthy but monotonous. This approach also reduces anxiety. Readers do not need to perform nutritional excellence at every meal. They need to widen variety, increase fiber where tolerated, and do so in ways they can maintain. Healthier microbiome habits become durable only when they fit into real life.

Diagram 11.2. The Weekly Plant Diversity Tracker

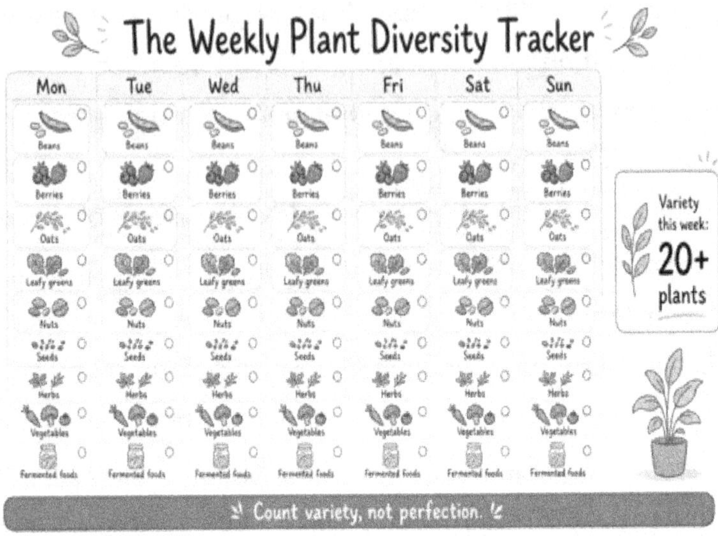

Stress, Sleep, and Digestive Consequences

By the time readers arrive at the microbiome section, many have already felt the gut-brain axis firsthand. Stress changes appetite, motility, bowel pattern, and symptom perception. Poor sleep alters hunger hormones, inflammatory tone, energy for food preparation, and the likelihood of relying on the most convenient available calories. The result is not a tidy one-way pathway but a loop: stress worsens sleep, poor sleep narrows choices, narrowed choices worsen digestive stability, and symptoms create more stress. A practical microbiome chapter must therefore speak about sleep and stress not as lifestyle garnish but as part of the gut environment itself.

This matters especially for readers who have begun to moralize their symptoms. When sleep is poor, and life is disordered, people often assume they need the correct supplement. More often they need a calmer sequence: enough rest to reduce reactivity, enough structure to eat predictably, and enough patience to distinguish temporary turbulence from a deeper problem. The microbiome is biologically complex, but for many readers the first useful intervention is simple rhythm.

Diagram 11.3. Stress, Sleep, and the Gut

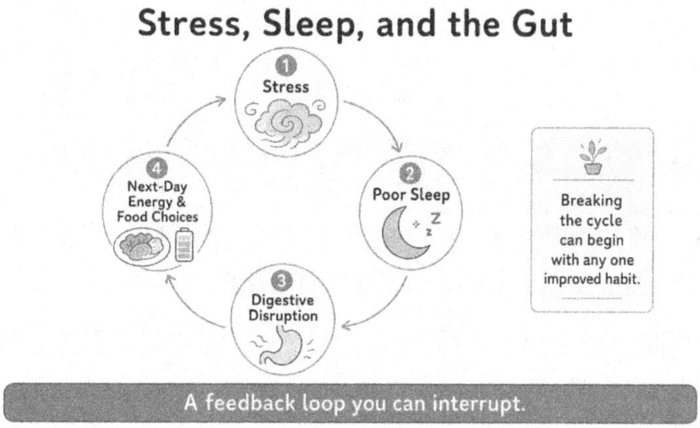

Travel, Alcohol, and Disruption

Healthy-person framing also matters when life stops being ideal. Travel changes meal timing, hydration, stress level, sleep, and exposure to unfamiliar foods. Alcohol may displace sleep quality and irritate the gut, especially at higher doses and frequency. Illness and antibiotics can temporarily flatten routine altogether. None of this means a person has failed at microbiome care. It means the microbiome, like the rest of physiology, is sensitive to disruption and usually responds better to time and consistency than to punishment.

Diagram 11.4. Alcohol on a Spectrum

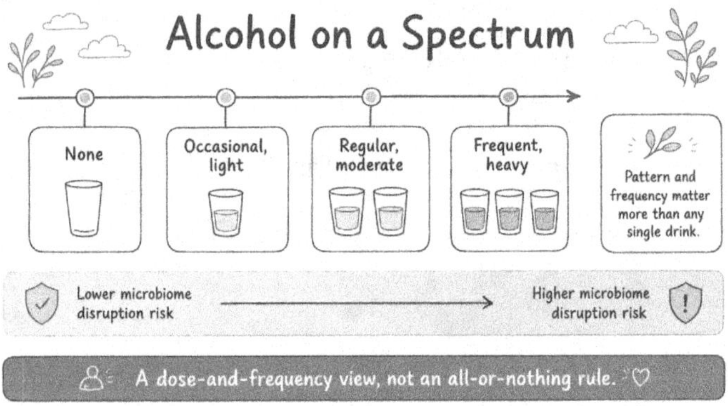

Read this chapter as permission to recover sensibly rather than react dramatically. After travel, the goal is not detoxification. After heavy social weekends, it is not self-reproach. After illness, it is not aggressive self-experimentation. The healthier path is re-entry into supportive basics: hydration, gentle meals, fiber progression as tolerated, movement when strength returns, and sleep. The body often recovers best when the reader stops trying to force the process.

Diagram 11.5. Travel and Recovery

A More Realistic Standard

The most health-promoting message this book can give many readers is not that they should optimize their microbiome. It is that they should stop thinking in all-or-nothing terms. A microbiome-supportive life is rarely a pure one. It is a life in which ordinary habits are tilted, repeatedly and imperfectly, toward better substrate, better rhythm, and better recovery. That is both less exciting than the marketing story and far more empowering.

Diagram 11.6. The Illness-Recovery Ladder

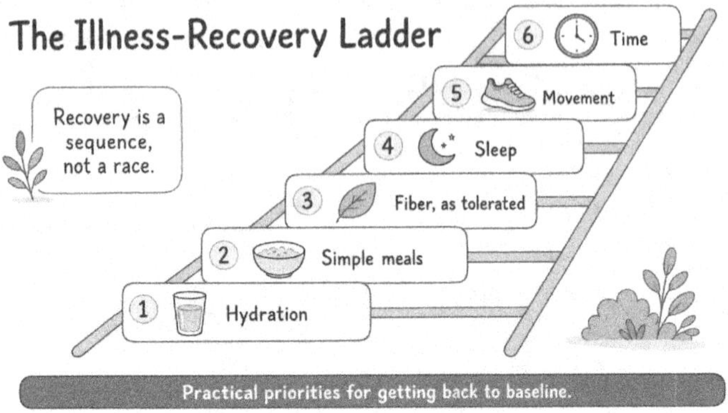

.

What follows from that is a humane standard. Good choices count. Better choices count more. Best choices are useful when available but are not the entry fee for health. Readers become healthier not by becoming purer, but by becoming more consistent. That is the daily-life lesson the microbiome field most needs to communicate.

Diagram 11.7. Good, Better, Best Daily Choices

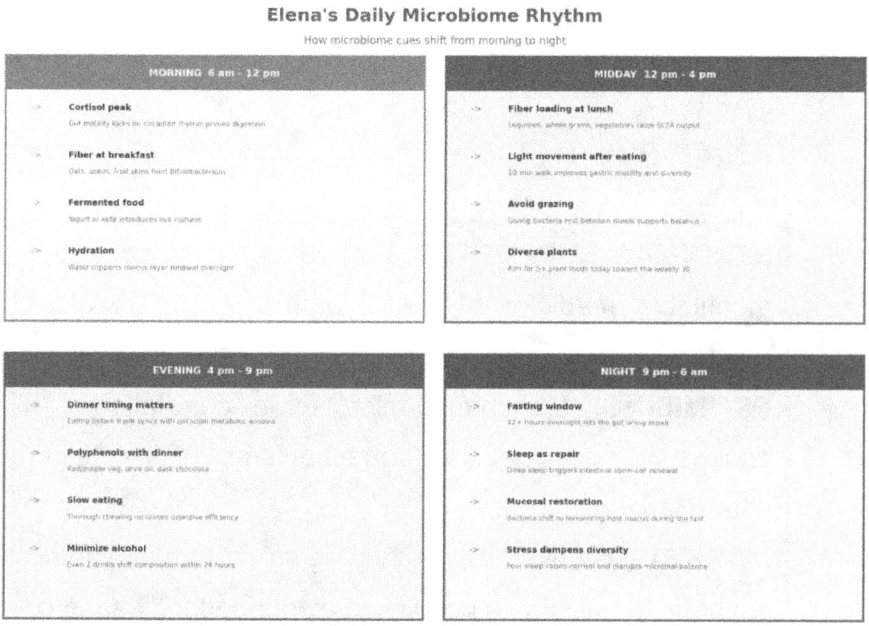

Diagram 11.7 | Daily rhythms reflect circadian biology, not rigid rules | Consistency matters more than perfection

Takeaways

- Your microbiome experiences life as repeated patterns, not isolated heroic choices.
- Weekly diversity, regular sleep, movement, and recovery routines usually matter more than searching for a single perfect product.
- Travel, stress, alcohol, illness, and disruption call for steady recovery, not panic.
- The healthiest microbiome advice is realistic enough to survive ordinary life.

12 The Microbiome Across the Decades

STRONGLY SUPPORTED: Diet variety is the strongest modifiable predictor of microbiome diversity in older adults.
PROMISING: Menopause-microbiome correlations — consistent associations, mechanisms still being characterized.
MARKETING-ONLY: 'Anti-aging' microbiome supplement claims — no clinical evidence for lifespan or healthspan benefit.

On a Sunday afternoon, four people from the same family are in the same house but living in very different microbial worlds. There is a toddler who still puts everything into her mouth. A teenager is running on short sleep and convenience food. A pregnant adult is learning that her body is changing in ways she can feel but not fully predict. And there is a grandmother whose appetite has narrowed and whose medication list has grown longer over the past decade. It is tempting to speak of "the microbiome" as if it were a single, stable entity. In lived reality, it changes across the lifespan, shaped by development, hormones, medications, diet, movement, illness, and the social conditions that make healthy routines easier or harder to sustain.

That lifespan view matters because broad health advice can quickly become vague. Readers need to know not only what helps a microbiome in principle, but what matters most at a given life stage. The priorities of infancy are not the priorities of adolescence. The questions of midlife are not the same as those of older adulthood. A more useful health book therefore asks a practical question at every stage: what are the most important microbiome-supportive conditions now?

Early Life: Seeding Is the Start, Not the Verdict

The microbiome story begins before memory. Birth mode, feeding patterns, skin contact, antibiotics, household environment, and early diet all influence the first microbial communities that colonize an infant. This is scientifically important, but it has too often been translated into parental guilt. That translation is both unfair and biologically simplistic. Early seeding matters, yet it is the beginning of a developmental trajectory rather than a final sentence. Children are shaped by a stream of exposures and interactions over time, not by one irrevocable opening event

Diagram 12.1. The Lifespan Arc

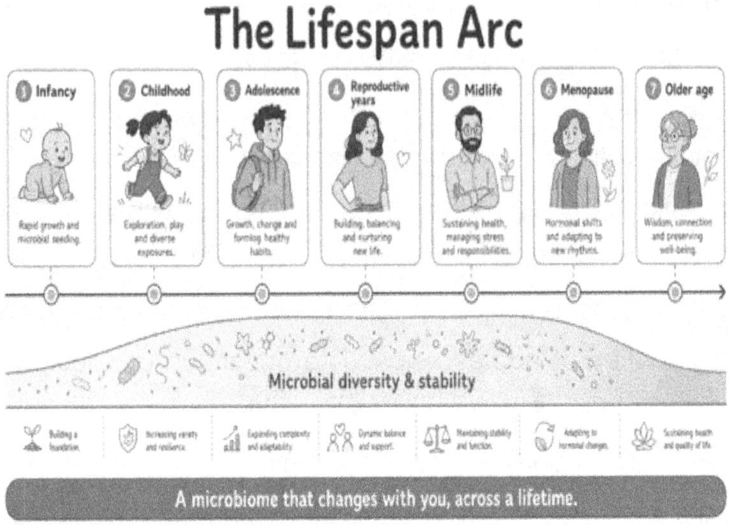

This is why the right tone for early-life microbiome discussion is supportive rather than moralizing. Parents should understand the factors that influence microbial development, but they should not be invited to treat infancy as a contest of perfection. When medically necessary interventions occur, they should be understood in that context. The useful question is how to support healthy development in the future: varied foods when appropriate, sensible antibiotic stewardship, time spent outdoors and with caregivers, and avoiding the temptation to turn every early-life variable into a referendum on future health.

Diagram 12.2. Early Seeding Pathways

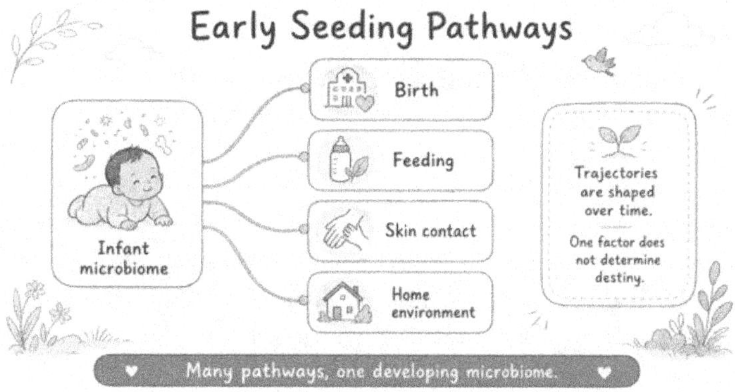

Adolescence and Reproductive Years

Adolescence is microbiome life under pressure. Sleep shortens, food quality often becomes irregular, stress rises, autonomy increases, and antibiotics may still enter the picture for common infections or acne-related treatment. It is also a period when habits become encoded in ways that can persist for years. The practical value of microbiome thinking in adolescence is therefore not in creating fear but in naming leverage: regular sleep when possible, less nutritional monotony, fewer ultra-processed defaults crowding out everything else, and a better understanding that how the body feels is often related to rhythm as much as to ingredients.

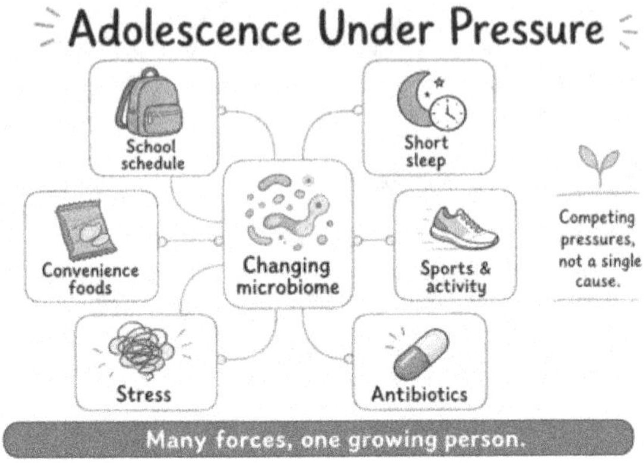

The reproductive years introduce another layer of complexity. Hormonal cycling, pregnancy, postpartum recovery, caregiving strain, and work-life compression all affect eating patterns, sleep, and microbiome stability. Readers in this stage often need relief from the pressure to do everything perfectly. A more helpful approach is to identify the highest-yield basics: adequate nourishment, fiber variety where tolerated, restoration after birth or illness, and a refusal to confuse social pressure with biological necessity.

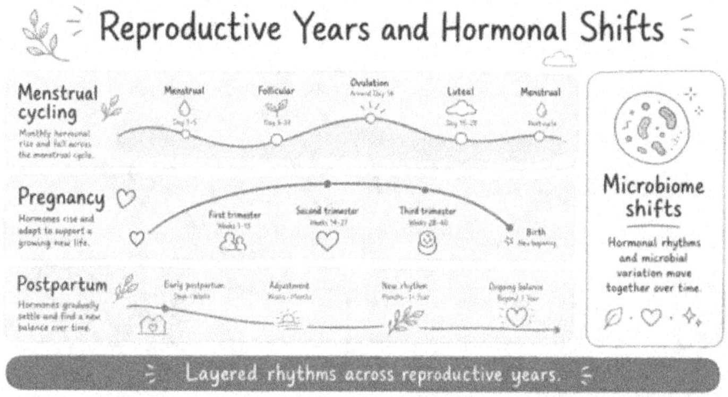

Midlife, Menopause, and Older Adulthood

Midlife frequently brings a collision of responsibilities. Sleep may worsen. Stress may become chronic. Exercise competes with work and caregiving. Weight, blood sugar, and inflammation become more visible concerns. The microbiome does not stand outside those pressures; it participates in them. This is one reason midlife readers often benefit from the least glamorous advice in the book: more routine, more plant diversity, more attention to alcohol frequency, more realistic stress recovery, and fewer oscillations between intense self-improvement and total exhaustion.

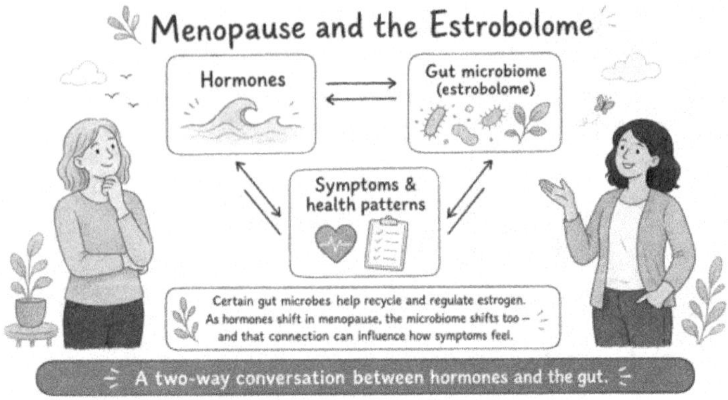

Menopause and older adulthood raise additional questions about hormones, appetite, medication burden, frailty risk, and social context. The emerging science of the estrobolome is one example of how gut microbes may interact with hormone metabolism. But for readers, the more immediate lesson is practical. In older age, microbiome health is often protected not by novelty but by preserving appetite, food quality, mobility, strength, hydration, and social connection. Isolation, polypharmacy, and low diet variety can shrink the conditions in which microbial resilience is maintained.

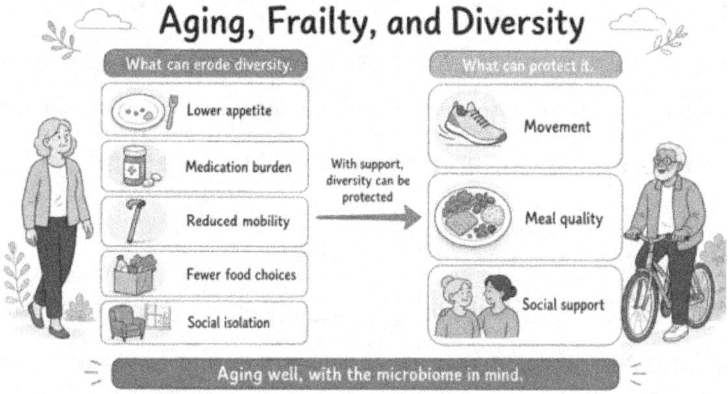

Harold's seventh decade is, by current research standards, the cohort with the thinnest longitudinal microbiome data. Most large-scale studies have followed adults through midlife; far fewer have followed the same individuals from sixty-five into the eighties and nineties. What is known is that diversity tends to narrow somewhat in older adulthood, that the strongest predictor of a 'younger-looking' microbiome in older adults is dietary variety rather than supplementation, and that frailty correlates with microbial monotony more than with any specific missing organism. Harold's practical takeaway is one his ninth-grade biology students would recognize: when the data is thin, the responsible move is to act on the principles the data does support and to stay curious about what the next decade of studies will show.

Leela, the infant from Chapter 2 whose microbiome report bewildered her mother, will reach adulthood in a decade where longitudinal microbiome studies will likely have caught up with her cohort. The data thin for older adults precisely because few studies have tracked the same individuals from infancy into old age. Leela's generation may be the first for whom that story can be told in full.

What Matters Most at Each Stage

Diagram 12.7. The Best High-Yield Action by Decade

A lifespan framework is valuable because it narrows the field. Readers do not need a thousand microbiome facts at every age. They need the few that change behavior. In infancy, the emphasis is trajectory and support. In adolescence, rhythm and food environment matter disproportionately. In adulthood, stress, time pressure, and the risk of chronic disease move to the foreground. In older adulthood, maintaining intake, movement, and resilience becomes central. Each stage has its own microbial logic, but the deeper theme is continuity: healthier conditions, repeated over time, matter more than dramatic

interventions imposed briefly

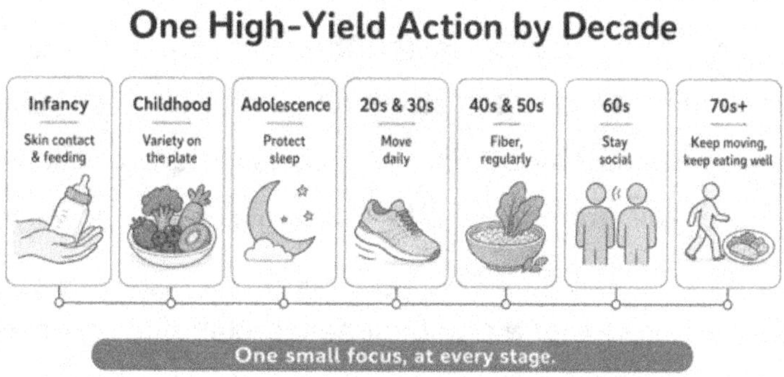

.

Takeaways

- The microbiome changes across the lifespan, so health advice should change with it.
- Early-life influences matter, but they are not destiny.
- Adolescence, reproductive years, midlife, and older adulthood each come with different high-yield priorities.
- The most useful microbiome guidance is age-aware, practical, and nonjudgmental.

13 A Personal Health Plan That Respects the Evidence

By the time a reader finishes this book, they may feel two opposing impulses at once. The first is hope: the microbiome clearly matters, and many daily habits that support it also support overall health. The second is uncertainty: the evidence is real but uneven, and many marketed interventions are still much louder than the data supporting them. A good final chapter should not resolve that tension by pretending it is gone. It should show the reader how to live intelligently inside it.

That is the purpose of a personal health plan that respects the evidence. This is not a protocol. It is not a stack of supplements, a home testing ritual, or a new reason to supervise every bowel movement with a spreadsheet. It is a way of making decisions that is calm, proportionate, and durable. The goal is to become a healthier person, not a more anxious patient.

Start With the Evidence Tier

Throughout this book, the most important intellectual tool has been the distinction among Strongly Supported, Promising, and Marketing-Only. That distinction belongs in your personal plan because it helps you decide what deserves effort, money, and emotional energy. Strongly Supported actions are the basics that help many systems at once: diet quality, fiber variety where tolerated, sleep regularity, physical activity, recovery after illness, and partnership with a clinician when symptoms are persistent or concerning. Promising interventions may be worth watching or discussing, but they should not become the center of your identity. Marketing-Only claims belong in the category of disciplined skepticism.

Using the evidence tier as a filter protects readers from a common trap: spending the most attention on the least reliable claims because they are the most exciting. The harder, healthier discipline is the opposite. Give the greatest consistency to the interventions with the strongest support, and reserve the greatest skepticism for the interventions making the loudest promises.

Diagram 13.1. The Evidence Filter in Real Life

Track Patterns Without Becoming Captive to Them

Self-observation can be useful. Many readers benefit from noticing how symptoms change when sleep collapses, travel accumulates, fiber intake changes abruptly, or stress becomes relentless. But tracking can also become its own disease, especially for readers already prone to health anxiety. The best tracking systems are brief, sustainable, and interpretive rather than obsessive. They help a person see patterns. They do not ask a person to become a full-time data analyst of their own body.

A useful rule is to track only what might change a decision. Meal patterns, sleep quality, bowel rhythm, medication changes, major stressors, and persistent symptoms can all be relevant. Minute-by-minute micromanagement usually is not. The point of awareness is better judgment. If tracking increases dread, narrows life, or drives impulsive product purchases, it has stopped serving health.

Diagram 13.2. Track Patterns, Not Noise

Know When to Try, and When to Call

One of the most empowering things a health book can do is teach the difference between a reasonable self-directed adjustment and a situation that needs clinical evaluation. A person with mild routine irregularity may reasonably work on meal timing, fiber progression, hydration, sleep, and movement. A person with blood in the stool, significant weight loss, persistent severe pain, fever, recurrent vomiting, or medication complications needs something different: medical assessment. Microbiome thinking should improve safety, not displace it.

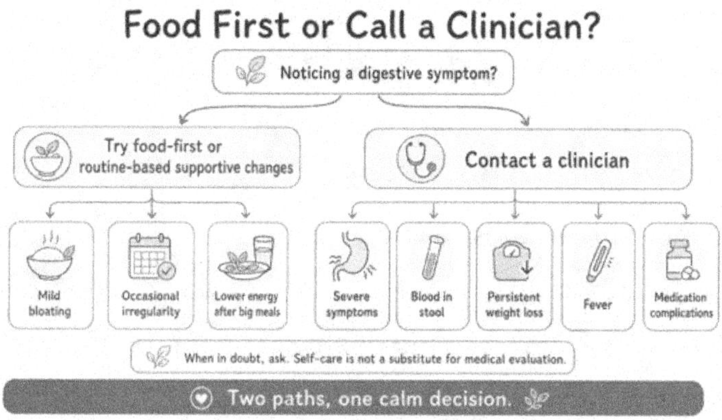

The same principle applies to commercial tests and products. A stool report may be interesting, but it is not the same as a diagnosis. A supplement label may sound scientific, but it is not the same as replicated human evidence. Personal responsibility in health does not mean doing everything alone. Often it means knowing when not to.

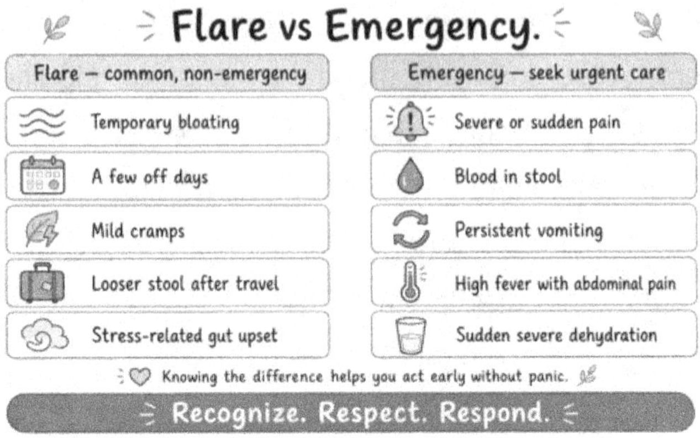

Build a Plan You Can Still Follow on a Bad Week

The personal plan most likely to improve health is not the most ambitious one. It is the one still standing when work is chaotic, family demands rise, sleep is poor, or motivation is gone. This is where many wellness plans fail. They are designed for a fantasy life and collapse under real pressure. A microbiome-aware plan should be simpler than your aspiration and sturdier than your mood.

For most readers, a realistic plan includes a short list: gradually broaden plant diversity, anchor meal timing where possible, protect sleep, move most days, recover rather than punish after disruption, and seek clinician support when symptoms cross clear thresholds. It may also include a 30-day experiment with one or two changes rather than seven. Health improves when change is organized enough to be repeated and modest enough to be believed.

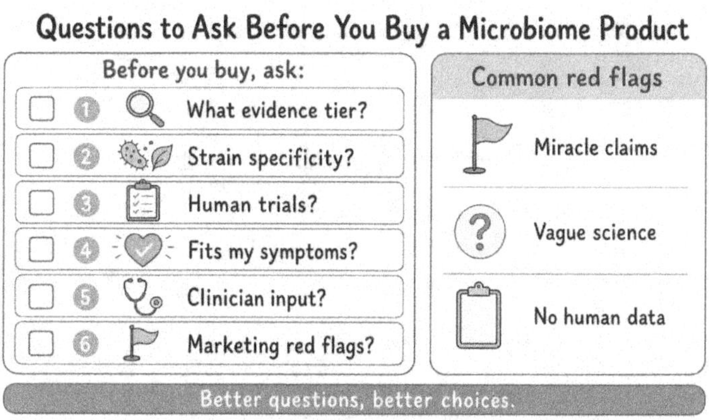

The Healthier Person at the End of the Book

If this book has done its job, the reader at the end is not someone who now believes the microbiome explains everything. It is someone who understands where the science is solid, where it is suggestive, and where it is being exploited. That reader is less vulnerable to hype, more confident in ordinary habits, and better able to distinguish curiosity from compulsion. They know that health is usually built from repeated conditions, not theatrical interventions.

This is a modest ending, but it is an honest one. The microbiome is powerful, complicated, and still partly mysterious. The wisest response is not worship or dismissal. It is informed partnership: with your body, with evidence, and with the routines that make better health more likely.

Diagram 13.7. The Healthy-Person Roadmap

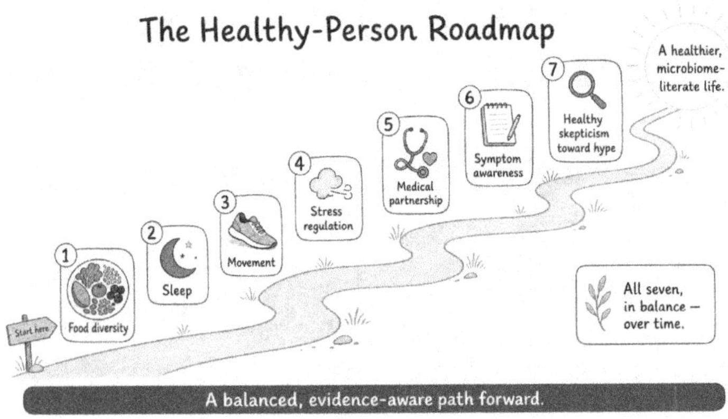

Return to Priya. She is now forty-three, and the colorful stool report that opened this book is in a folder she rarely consults. What changed in her first year of reading is not her IBS diagnosis — still active, still managed — but the questions she brings to it. She has stopped buying probiotics that promise to fix her gut. She has started keeping the same handful of fermented foods in the refrigerator week to week. She has worked with a registered dietitian on a personal trigger list rather than a universal elimination diet. She has had one conversation with her gastroenterologist in which she used the phrase 'low-evidence claim' without irony, and her doctor smiled because she had earned it. She has also, on a recent bad week, missed every single thing on her plan and resumed the next Monday without treating the lapse as a verdict. None of this looks like a transformation from the outside. It looks like a person who reads the news about her own condition with calmer eyes. That is what the book was trying to make possible.

Harold, the retired teacher, has done his version of this exercise. His plan is not a regimen. It is a short list taped to the inside of a kitchen cabinet: aim for thirty different plant foods most weeks, walk after dinner three or four nights, keep alcohol to two evenings a week rather than five, talk to his physician before starting any new supplement his grandchildren recommend. He has resisted the urge to track anything in an app. He has not bought a microbiome test. He has decided that the citizen-science sample he mailed in last spring is participation enough. What he has built is the thing this chapter has been trying to describe: a plan sturdy enough for a bad week, modest enough to last a decade, and his own in a way no product could be.

Takeaways

- Use the book's evidence tiers to decide what deserves trust, effort, and money.
- Track patterns only to the extent that they improve judgment and reduce confusion.

- Know the difference between self-care, product marketing, and symptoms that require medical evaluation.
- Build a plan sturdy enough to survive ordinary life; that is the plan most likely to improve health.

The Closing Call: Stay Curious, Stay Skeptical

You began this book, perhaps, with a general sense that gut bacteria mattered and that you should be doing something about them. What you have now is different. You have a framework, a set of tools for thinking about this field rather than just a set of facts about it.

That framework has three pillars. The first is genuine curiosity. The microbiome is an astonishing system, and the science being done to understand it is some of the most creative and consequential biology of our era. Letting yourself be genuinely engaged by that is appropriate, not naive. The second pillar is disciplined skepticism: the kind of principled questioning that asks what the evidence is, how large the study was, whether it was in humans or only in mice, whether the effect replicated, and whether the person making the claim is trying to sell something. Curiosity without skepticism is gullibility. Skepticism without curiosity is sterility. Together they produce literacy.

The third pillar is sustained engagement. Microbiome science is not finished. Research will continue to arrive, and some of it will confirm what this book has described, while some will complicate, qualify, or overturn earlier conclusions. A treatment that seems promising today may fail in Phase III trials. New species will be identified, new mechanisms described, and new connections drawn between the microbiome and conditions we have not yet considered investigating. Each of those developments is an invitation to update your understanding.

Richard finished his physical with a referral to a dietitian specializing in gut health and a handout from his physician about the 30-plant-foods goal. He learned that his bloating was likely due to a drop in dietary fiber after retirement, when the daily salad he ate at work was replaced by convenience foods at home. He declined the probiotic his consumer test had recommended, having established that it contained no validated strains for his situation. Instead, he started making a fermented vegetable once a week, increased the variety of legumes and whole grains in his diet, and added an evening walk to his morning walk with the dog. He returned to his doctor six months later feeling noticeably better. He also signed up for a university gut health study in his city. His doctor was glad he had brought his printout to the appointment.

Tools and Habits: Your Daily Microbiome Practice

Microbiome literacy is partly intellectual and partly practical. The practical side is what you do on an ordinary Tuesday. The interventions with the strongest evidence on ordinary Tuesdays are not tests, supplements, or sophisticated interventions. They are the unglamorous daily habits that add up over months and years to measurable differences in microbial diversity, short-chain fatty acid production, and the downstream benefits those bring.

The thirty-plants-per-week target is more achievable than it sounds once you count every distinct variety of vegetable, fruit, whole grain, legume, nut, seed, herb, and spice. A salad with five different vegetables, walnuts, sunflower seeds, and an oregano-lemon dressing already counts as nine plant foods. A mixed-grain porridge with banana, berries, and flaxseed is five more. A dinner of lentil soup with cumin, coriander, bay leaf, garlic, and diced tomatoes adds another six. You are not searching for exotic ingredients. You are counting the diversity already present in your kitchen with more deliberate attention.

Sleep is the most undervalued variable in microbiome health discussions. Diet is the single most powerful modifiable driver of microbiome composition. Still, sleep disruption alters the microbiome through multiple pathways: changes in cortisol rhythms, reduced gut motility, and shifts in the timing of microbial metabolic activity, which is normally synchronized to your circadian cycle. People who sleep fewer than six hours per night show measurable reductions in microbial diversity in multiple studies. Prioritizing sleep is one of the most evidence-based microbiome interventions available.

Exercise matters through mechanisms beyond burning calories or improving cardiovascular fitness. Physically active people show greater microbiome diversity and higher levels of short-chain fatty acid production than their sedentary counterparts, even when diet is held roughly constant. The mechanisms include increased colonic transit time, favorable changes in intestinal oxygen levels that support anaerobic bacteria, and exercise-induced changes in bile acid production. A thirty-minute walk most days of the week produces measurable effects.

Managing stress is listed on the daily toolkit, and the science behind it is real. Chronic stress elevates cortisol, increases intestinal permeability, alters gut motility, and directly influences which bacterial populations thrive in the gut environment. These effects contribute to the well-documented worsening of IBS symptoms under stress and to the gut microbiome disruptions seen in people with chronic anxiety or depression. The interventions with the most evidence for stress-related microbiome modulation include mindfulness-based stress reduction, regular aerobic exercise, consistent sleep scheduling, and social connection. All of them are worth building into daily life.

Takeaways

- The microbiome evidence hierarchy sorts recommendations into three tiers: strongly supported lifestyle foundations, promising interventions with emerging but incomplete human evidence, and a broad third tier of commercial products that lack the clinical evidence to justify their claims. Know where a recommendation sits before acting on it.
- Consumer microbiome tests provide real data, but interpreting them requires understanding what 16S sequencing does and does not measure, what diversity scores do and do not mean at the individual level, and why a company's product recommendations deserve independent verification before you spend money on them.
- Productive conversations with your physician about gut health are possible right now, without any new technology. The most valuable questions involve antibiotic stewardship, condition-specific dietary evidence, and whether microbiome-informed care options are relevant to your situation.
- The maturation of live biotherapeutic products will define the next decade of microbiome medicine, the integration of microbiome profiling into cancer treatment, and the gradual incorporation of microbiome

monitoring into routine preventive care. Each of these is real and in progress; none of them is fully here yet.

- Equity in microbiome medicine requires deliberate action at the level of clinical trial enrollment, insurance coverage advocacy, reference database diversity, and structural determinants of the lifestyle behaviors that most powerfully shape the microbiome. These are decisions being made now, and informed citizens can influence them.
- Citizen science projects including the American Gut Project, ZOE, the Earth Microbiome Project, and MetaSUB offer genuine opportunities to contribute your data to the research base that will shape the next generation of microbiome medicine. Participation is accessible, meaningful, and consequential.
- Microbiome literacy is a practice, not a destination. The science will continue to evolve. Your role is to stay curious enough to engage with it, skeptical enough to evaluate it critically, and engaged enough to advocate for its application in ways that serve everyone, not only those with the most resources to access it.

14 Closing

You arrived at this book with a question you could not quite answer on your own. Maybe a stool test report arrived in the mail full of bacterial names that meant nothing. Maybe a headline about fecal transplants left you fascinated and uncertain. Whatever the doorway, you walked through it. Having traveled thirteen chapters through one of the most alive frontiers in modern science, you are not the same reader who opened the first page. The knowledge you have built is not trivia but a framework, because the field will keep moving, and you now have the tools to keep pace with it.

The Journey Behind You

Think back to Chapter 1. Metagenomics has given science a new pair of eyes, revealing a living universe hiding in plain sight inside every human body, every gram of soil, every cup of seawater. Chapter 2 gave that universe a neighborhood map: the gut, the skin, the oral cavity, and the urogenital tract each maintain their own distinct ecology. Chapter 3 took you behind the laboratory door, showing how a stool sample becomes a consumer report and why understanding the limits of sequencing technology sharpens your ability to read claims.

Chapter 4 made the science personal. The gut-brain axis is not a metaphor. The vagus nerve, microbial production of serotonin and GABA, and the mouse personality studies all point toward a real bidirectional conversation between gut and brain that has been ongoing for as long as you have been alive. Chapter 5 traced how birth mode, breastfeeding, and early antibiotic exposure shape immune trajectories that persist for decades. When the microbial richness of early life is disrupted, the costs are real.

Chapter 6 gave you the book's most actionable findings. Plant variety, dietary fiber, fermented foods, and polyphenols are the substrates your gut community depends on. The Weizmann Institute's personalized nutrition research showed that the microbiome is one of the strongest predictors of individual response to food. Chapter 7 held the book's most necessary tension: antibiotics are the most life-saving medicines ever developed, and when overused, they drive the microbiome disruption linked to chronic disease. Antibiotic stewardship is pro-medicine at the level of the whole ecosystem.

Chapter 8 gave you an honest ledger: FMT with strong clinical evidence for recurrent C. diff, modest strain-specific support for certain probiotics, and a market of products that have outrun the science. Chapter 9 pulled the camera back to a planetary view, where the same sequencing technology maps immunotherapy responses and monitors the ocean microbiomes that produce half of Earth's atmospheric oxygen. Chapter 10 distilled everything into the practical: a hierarchy of what the evidence supports, a guide to reading test results critically, and an honest look at what comes next.

The Next Ten Years

Standing at the edge of what is known, the picture holds extraordinary possibility alongside real work still to be done.

Better diagnostics are the most concrete near-term change. Today's consumer microbiome tests have outpaced the interpretive science behind them. As validation studies accumulate and reference databases grow more diverse, that gap will close. A microbiome panel with the clinical validity of a cholesterol test could be achieved before 2036. The FDA approvals of Rebyota and Vowst for C. diff in 2022 and 2023 proved that defined microbial communities can be regulated, manufactured, and delivered as drugs. Trials are active for inflammatory bowel disease, metabolic syndrome, and microbiome-guided cancer immunotherapy.

Microbiome medicine left to market forces alone will reach affluent populations first. The patients who most need it are often least likely to have access to testing or the physician relationship that allows for nuanced conversations. Advocacy, policy attention, and diversification of clinical trial populations are the mechanisms that change this trajectory. Readers who care about the science play a real role in advancing those changes.

Citizen science will also grow, and the growing will matter. The American Gut Project, the ZOE Health Study, and the MetaSUB consortium have shown that ordinary people contributing samples and data make the research more powerful and more representative. Your data, carefully contributed to a vetted project, is a genuine addition to the body of knowledge that clinical medicine will draw on.

You as a Participant

There is a tendency, when reading science, to experience it as something that happens elsewhere, in laboratories, on the pages of journals you will never read. This book has argued, from its first page to this one, that this tendency is mistaken.

You are an active participant in microbiome science. Every informed conversation with a physician about antibiotic stewardship contributes to better prescribing culture. Every precise question you bring to a medical appointment shifts the quality of that exchange. Every time you share a grounded explanation of what probiotics can and cannot do with a friend about to buy an expensive supplement based on influencer advice, you are doing public health communication where it most often counts.

This framework is not a prescription but an invitation. Asking better questions of your healthcare providers requires only that you bring this book's questions into rooms where they can be acted on. Supporting the science can mean participating in a citizen science study or sharing verified information rather than viral health claims. Voting with your diet means nourishing the ecosystem as a whole. Sharing your knowledge means treating what you have learned as something worth passing on. Staying open to new evidence matters most of all, because the field is young, revisions are real, and the willingness to update when the data changes is the mark of scientific literacy.

A Personal Note

I want to be honest about something I felt throughout writing this book: genuine excitement mixed with a kind of responsible discomfort. The excitement is easy to explain. That a fecal transplant can cure a recurrent infection that months of antibiotics could not, that gut bacteria produce most of the body's serotonin, that microbial diversity in a teaspoon of healthy soil exceeds the number of people on Earth: all of it carries the sense of vertigo and wonder that I imagine early astronomers felt when they first grasped the scale of the universe.

The discomfort comes from translating genuinely uncertain science for readers hoping for clear answers. There are places where the science is strong, and the message is clear. There are others in which the animal data is compelling, and the human clinical evidence remains sparse. I have tried, every time, to tell you which is which, because honest uncertainty is more useful than false confidence.

What I hope you carry forward is not a list of recommendations but a posture: curious, skeptical, humble, and lit up by the complexity of the living world. The microbiome within you is not a problem to be optimized. It is a community, ancient and responsive, part of the human story for hundreds of thousands of years. Learning to understand it is a grounded form of care for the remarkable ecosystem you carry everywhere you go.

Stay curious. The best findings are still ahead.

---

15 Glossary of Key Terms

This glossary defines the most important scientific terms used throughout the book. Each entry is written in plain language so that readers with no prior background in biology can follow along. If a word in the main chapters felt unfamiliar, you will find it explained here without jargon.

16S rRNA Sequencing

The 16S ribosomal RNA gene is a short stretch of DNA found in nearly every bacterium on Earth, yet varies just enough between species to serve as a fingerprint. When scientists want to identify which bacteria are present in a gut sample, they extract all the DNA, amplify a specific gene, and then read its sequence using modern sequencing machines. By comparing those sequences against large reference databases, researchers can catalog the types and proportions of bacteria present without ever needing to grow them in a lab dish. This method transformed microbiome science because most gut bacteria cannot survive in oxygen-rich lab conditions, making traditional culture methods nearly useless for studying them. The technique has some limitations, particularly around distinguishing closely related species or strains. Still, it remains one of the most widely used entry points into understanding who lives in the gut.

Akkermansia muciniphila

Akkermansia muciniphila is a bacterium that makes its home in the mucus layer lining the gut wall, feeding on the mucin proteins produced by intestinal cells. Despite living on what the body secretes, it is associated with a healthy, thick mucus barrier rather than a depleted one, as it appears to stimulate the gut to replenish mucin continuously. Research has linked higher Akkermansia levels with lower obesity rates, better blood sugar control, and reduced gut inflammation. When Akkermansia levels drop, as can happen with a diet low in diverse plant foods or after antibiotic use, the mucus layer may thin and become more permeable. Scientists are actively studying whether this bacterium can be used as a next-generation probiotic. However, at the time of writing, most products on the market contain strains that have been studied far more than Akkermansia.

Bacteroidetes

Bacteroidetes is one of the two dominant phyla, or major groupings, of bacteria found in the human gut. The phylum includes hundreds of species, many of which specialize in breaking down complex carbohydrates that the human body cannot digest on its own. Members of the Bacteroidetes group tend to thrive on dietary fiber and plant-based polysaccharides, and they are among the primary producers of short-chain fatty acids in the colon. The relative balance between Bacteroidetes and Firmicutes, the other dominant phylum, has been studied extensively in relation to body weight and metabolism. However, the science is more complicated than early headlines suggested. A healthy, fiber-rich diet generally supports a robust Bacteroidetes population, which in turn contributes to stable energy metabolism and a well-nourished gut lining.

Bile Acids

Bile acids are molecules made in the liver from cholesterol and secreted into the small intestine to help digest fats and fat-soluble vitamins. After doing their digestive work, most bile acids are reabsorbed in the lower small intestine and recycled back to the liver in a loop called the enterohepatic circulation. The gut microbiome plays a central role in modifying bile acids along the way. Bacteria transform the primary bile acids produced by the liver into dozens of secondary bile acid variants, each with distinct biological effects on the gut lining, the immune system, and even the liver itself. Disruptions to this bile acid transformation process have been linked to conditions ranging from irritable bowel syndrome and inflammatory bowel disease to certain liver disorders. Because bile acids also act as signaling molecules that influence the brain and metabolic organs, they represent one of the key chemical bridges through which gut bacteria affect the rest of the body.

Commensal

A commensal organism lives alongside a host and benefits from the relationship without causing obvious harm. In the gut, the vast majority of the trillions of bacteria present are considered commensals: they take up residence, consume food that passes through, and in return contribute to digestion, immune education, and protection against harmful invaders. The term comes from a Latin phrase meaning "sharing a table," which captures the idea of organisms benefiting from the same food source. In practice, the line between commensal and mutualist is blurry, because many bacteria once labeled as neutral commensals turn out to provide active benefits when studied more closely. The distinction matters clinically because some commensals can become opportunistic pathogens if they grow out of control or enter tissues where they do not belong, a process sometimes called pathobiont activation.

Dysbiosis

Dysbiosis refers to an imbalance in the composition or function of the gut microbiota. It does not describe a single pattern but rather a broad departure from the community structure associated with good health, which can mean reduced diversity, overgrowth of potentially harmful species, loss of beneficial bacteria, or a shift in the types of metabolites the community produces. Dysbiosis has been linked to a wide range of conditions, including inflammatory bowel disease, irritable bowel syndrome, obesity, type 2 diabetes, mood disorders, and autoimmune diseases. Because the definition of a "healthy" microbiome is still being refined and varies between individuals and populations, dysbiosis is more of a conceptual framework than a precisely diagnosed state. Identifying dysbiosis in a clinical setting is an active area of research, and current stool tests sold directly to consumers have significant limitations in detecting or characterizing it.

Enterocyte

An enterocyte is a specialized absorptive cell that lines the wall of the small intestine. These cells are packed tightly together in a single layer that covers the enormous folded surface of the intestinal lining. Each enterocyte has a brush-border surface covered in tiny projections called microvilli, which dramatically increase the surface area available for absorbing nutrients such as sugars, amino acids, and fatty acids from digested food. Enterocytes are in direct contact with the contents of the gut lumen, including bacteria and their metabolites, thereby serving as a critical interface between the outside world and the bloodstream. They are replaced roughly every three to five days as new cells migrate up from the base of the intestinal crypts, providing a continuously refreshed barrier. Disruption of enterocyte function or integrity, as can happen in inflammatory conditions or with certain infections, compromises nutrient absorption and allows unwanted molecules to pass into tissues.

Fecal Microbiota Transplant

A fecal microbiota transplant, commonly abbreviated as FMT, is a procedure in which stool from a healthy, carefully screened donor is transferred into the recipient's gut. The goal is to introduce a diverse, balanced microbial community into a gut whose own ecosystem has been severely disrupted. FMT has become a standard treatment for recurrent infections with Clostridioides difficile, a dangerous gut bacterium that can take over after antibiotic therapy destroys the normal microbiome. In clinical trials for C. diff, FMT achieves cure rates that far exceed those of antibiotics alone. Researchers are now investigating FMT for other conditions, including inflammatory bowel disease, metabolic syndrome, and even certain neurological conditions, with variable results. The procedure is not without risks, including the possibility of transferring undetected pathogens from donor to recipient, which is why careful donor screening protocols are essential.

Firmicutes

Firmicutes is one of the largest and most diverse phyla of bacteria in the human gut, containing hundreds of species ranging from Lactobacillus strains associated with fermented foods to Clostridiales species that produce short-chain fatty acids from fiber. The phylum was once studied intensively in the context of obesity because early research suggested that people with obesity tended to have a higher Firmicutes-to-Bacteroidetes ratio than lean individuals. Subsequent research complicated this picture considerably, showing that the ratio varies widely across healthy individuals of different body types, diets, and geographic backgrounds, and that it is not a reliable diagnostic marker on its own. Many Firmicutes members are highly beneficial, particularly those that ferment fiber and produce butyrate, the main energy source for colonocytes. The diversity within this phylum is so vast that it is almost more useful to discuss individual species or genera than to speak of Firmicutes as a functional unit.

Gut-Brain Axis

The gut-brain axis is the two-way communication network linking the gastrointestinal tract and the central nervous system. It operates through multiple channels simultaneously, including the vagus nerve (a long cranial nerve that carries signals directly between the gut and the brainstem), the enteric nervous system (the hundreds of millions of nerve cells embedded in the gut wall, sometimes called the second brain), the immune system, and a flood of circulating hormones and metabolites produced by gut bacteria. The gut microbiome is a major participant in this network, producing neurotransmitters and neuroactive compounds, modifying the molecules that nerve endings detect, and influencing how the immune system communicates with the brain. Research in mice has shown dramatic behavioral changes when the microbiome is altered, and human studies have begun to reveal associations between gut bacteria and mood, anxiety, and cognitive function. However, the mechanisms in people are still being worked out.

Lactobacillus

Lactobacillus is a genus of bacteria named for their ability to produce lactic acid from sugars. These bacteria are among the most studied and widely recognized in gut and vaginal microbiome research. They are the microorganisms behind many fermented foods including yogurt, kefir, sauerkraut, and kimchi. In the vagina, Lactobacillus species are dominant in the microbiomes of many people and are associated with protection against infections by maintaining an acidic environment that discourages pathogen growth. In the gut, Lactobacillus species are present but not numerically dominant in most healthy adults; they are more abundant in the infant gut, especially in breastfed babies. Many commercial probiotic products contain Lactobacillus strains, and while some strains have demonstrated specific benefits in clinical trials for conditions such as antibiotic-associated diarrhea, these benefits are highly strain-specific and do not automatically apply to all Lactobacillus products.

Lipopolysaccharide

Lipopolysaccharide, often abbreviated as LPS, is a molecule found in the outer membrane of Gram-negative bacteria, which include many common gut residents. LPS is a potent signal to the mammalian immune system, triggering a strong inflammatory response when the immune cells that patrol the body encounter it. In a healthy gut with an intact mucosal barrier, LPS remains confined to the gut lumen and does not enter the bloodstream. However, when the gut barrier is compromised, small amounts of LPS can leak into circulation, a situation sometimes called endotoxemia. Chronic low-level endotoxemia has been associated with systemic inflammation, insulin resistance, and metabolic disease. Diets high in saturated fat appear to facilitate LPS absorption by packaging it into fat particles, a proposed mechanism linking high-fat dietary patterns to inflammation. Maintaining a healthy, intact gut barrier is therefore important not only for digestive reasons but also for managing the body's overall inflammatory state.

Metagenomics

Metagenomics is the study of all the genetic material collected from a community of organisms in a given environment, without first isolating or culturing individual species. In gut microbiome research, shotgun metagenomics means extracting all the DNA from a stool sample and sequencing it in large amounts at once, capturing fragments from every organism present, including bacteria, viruses, fungi, and archaea. This approach provides far more detail than 16S rRNA sequencing alone, because it sequences the entire genome of every microbe present rather than just one marker gene. Researchers can use metagenomics to identify specific species and strains, detect which genes are present, and infer the metabolic functions the whole community can perform. The computational challenge is enormous, because the data must be assembled and compared against reference databases containing billions of genetic sequences. Metagenomics has become the gold standard for detailed microbiome characterization in research settings.

Microbiome

The microbiome refers to the complete collection of microorganisms, their genes, and their metabolic products in a specific environment. When people talk about the human gut microbiome, they mean everything living in the gastrointestinal tract: bacteria, viruses, fungi, archaea, and other microscopic organisms along with the full complement of their genes and the chemicals they produce. The word microbiome is sometimes used interchangeably with microbiota, though technically microbiota refers to the organisms themselves while microbiome refers to the organisms plus their genetic content and products. The gut microbiome is by far the densest and most diverse microbial community in the human body. Still, people also have distinct microbiomes at sites including the skin, mouth, lungs, and urogenital tract. Because each person's gut microbiome is shaped by genetics, early-life exposures, diet, environment, and history of antibiotic use, no two people share an identical microbial community.

Microbiota

Microbiota is the collective noun for the microorganisms, primarily bacteria, that inhabit a specific environment such as the gut. The term is slightly more precise than "microbiome" when referring to the living organisms themselves, setting aside their genetic and biochemical outputs. The human gut microbiota is extraordinarily dense, particularly in the colon, where bacteria reach concentrations of hundreds of billions per milliliter. The community is not static. It shifts across the course of a day in response to meals, and it changes across a lifetime in response to aging, illness, medication, and changes in diet or geography. Studies comparing the gut microbiota of people living very different lifestyles, such as hunter-gatherer communities versus industrialized urban populations, reveal striking differences in diversity and composition, suggesting that many features of the modern microbiota may be departures from what the human gut evolved alongside over millions of years.

Mucin Layer

The mucin layer, also called the mucus layer, is the protective gel-like coating that lines the inside of the gastrointestinal tract. It is secreted by specialized goblet cells embedded in the intestinal epithelium and consists primarily of large glycoprotein molecules called mucins. The mucin layer serves multiple critical functions. It acts as a physical barrier that keeps most bacteria at a safe distance from the epithelial cells beneath, while allowing nutrients to pass through. It also harbors beneficial bacteria that live in the outer mucus layer and play roles in immune education and barrier maintenance. Akkermansia muciniphila, discussed above, is one of the best-known inhabitants of the mucus layer. When the mucus layer thins, whether due to a low-fiber diet, dysbiosis, or inflammatory conditions, bacteria can come into closer contact with the epithelium, potentially triggering immune reactions. A diet rich in diverse fibers appears to support goblet cell activity and mucus layer thickness.

Mycobiome

The mycobiome refers to the community of fungi that live in the gut alongside bacteria, viruses, and other microorganisms. Although fungi represent a small fraction of the gut microbiota by number, they are far more abundant than bacteria, and their metabolic influence can be significant. The gut mycobiome includes species from genera such as Candida, Saccharomyces, Malassezia, and Aspergillus. Most of these fungi are commensals under normal conditions, but some, particularly Candida albicans, can become problematic if they overgrow following antibiotic-driven disruption of the bacterial community. The mycobiome interacts directly with gut bacteria, and some fungi and bacteria engage in antagonistic relationships that help regulate each other's populations. Research into the mycobiome is considerably less advanced than research into the bacterial microbiome, partly because fungi are harder to culture and because reference databases for fungal genetics are less complete. Emerging evidence links mycobiome disruptions to inflammatory bowel disease and other gut conditions.

Postbiotic

A postbiotic is a preparation of inanimate microorganisms or their components that confers a health benefit on the host. The term distinguishes these preparations from probiotics (live microorganisms) and from prebiotics (food components that feed microbes). Postbiotics can include heat-killed bacteria, bacterial cell wall fragments, short-chain fatty acids produced by fermentation, or other metabolites isolated from microbial activity. Because they do not contain live organisms, postbiotics are more shelf-stable and easier to standardize than traditional probiotics. They may also be suitable for individuals who cannot safely consume live bacteria, such as severely immunocompromised patients. The science of postbiotics is rapidly developing, and some preparations have shown promise in supporting immune function and gut barrier integrity. Regulatory definitions vary by country, and the commercial market for postbiotic products has grown faster than the clinical evidence base supporting specific products.

Prebiotic

A prebiotic is a substrate that is selectively used by host microorganisms in a way that confers a health benefit. In practical terms, most prebiotics are dietary fibers or complex carbohydrates that humans cannot digest but that gut bacteria ferment readily. Classic examples include inulin, found in chicory root and Jerusalem artichokes; fructooligosaccharides, present in garlic and onions; and beta-glucans, found in oats and barley. When these fibers arrive in the colon intact, bacteria break them down, producing short-chain fatty acids as byproducts that nourish both the bacteria and the colonocytes. Feeding beneficial bacteria with prebiotic fibers can increase their abundance and diversity over time, and some studies show measurable improvements in gut barrier function, bowel regularity, and immune markers. The definition of prebiotic is deliberately broad and continues to evolve as researchers discover that compounds beyond fiber, including certain polyphenols and human milk oligosaccharides, also meet the criteria.

Probiotic

A probiotic is a live microorganism that, when administered in adequate amounts, confers a health benefit on the host. This official definition from the International Scientific Association for Probiotics and Prebiotics emphasizes two key requirements: the organism must be alive upon delivery, and the benefit must be demonstrated through scientific evidence. Common probiotic strains include specific Lactobacillus and Bifidobacterium varieties, as well as the yeast Saccharomyces boulardii. The evidence for probiotics is highly strain-specific and benefit-specific: a strain that demonstrably reduces antibiotic-associated diarrhea does not automatically prevent bloating or support weight loss. The commercial probiotic market is enormous, and many products make claims that exceed what the research supports for their particular strains or doses. Probiotics appear most effective in defined clinical scenarios, including specific gastrointestinal infections, gut protection in premature infants, and some cases of irritable bowel syndrome, rather than as a general wellness supplement for everyone.

Short-Chain Fatty Acids

Short-chain fatty acids, commonly abbreviated as SCFAs, are small organic acids produced when gut bacteria ferment dietary fiber and other non-digestible carbohydrates in the colon. The three most important in human health are acetate, propionate, and butyrate. Butyrate is the preferred fuel source for colonocytes, the cells lining the colon wall, and it also has anti-inflammatory properties and helps regulate cell growth and prevent excessive proliferation. Propionate travels to the liver, where it participates in glucose metabolism and signals fullness to the brain. Acetate enters the general circulation and contributes to energy metabolism throughout the body. SCFAs are among the primary mechanisms by which the gut microbiome influences systemic health. A diet rich in fermentable fibers supports robust SCFA production, while low-fiber Western diets tend to reduce SCFA levels and may contribute to a less-nourished gut lining.

Strain

In microbiology, a strain is a specific genetic variant within a species. Two bacteria can belong to the same species yet differ enough genetically to behave quite differently inside the body. This distinction is enormously important in microbiome science and particularly in the probiotic industry, where the relevant benefits are nearly always strain-specific. For example, Lactobacillus rhamnosus GG is one specific strain of the species Lactobacillus rhamnosus; it has been studied in hundreds of clinical trials and has well-documented benefits for certain conditions, but another strain of the same species might have completely different properties. When reading about microbiome research or evaluating probiotic products, the strain designation (usually given as a letter-number code after the species name) is the most clinically meaningful level of identification. Researchers use whole-genome sequencing and other molecular tools to characterize and compare strains with high precision.

Symbiosis

Symbiosis is a broad biological term describing a close and long-term relationship between two different species. The relationship can take several forms. Mutualism describes a relationship where both parties benefit, as when gut bacteria produce vitamins and short-chain fatty acids that nourish the host while receiving a warm, nutrient-rich environment in return. Commensalism describes a relationship in which one organism benefits while the other is unaffected. Parasitism occurs when one organism benefits at the expense of the other. The relationship between humans and their gut microbiota is best described as an evolving mutualism. However, it also contains elements of commensalism, because not every microorganism in the gut actively benefits its host. The gut microbiome is sometimes described as a co-evolved symbiotic community, meaning that humans and their microbial inhabitants have shaped one another over millions of years of shared evolution, resulting in deep biological interdependence.

Tight Junctions

Tight junctions are protein structures that connect neighboring epithelial cells in the intestinal wall, sealing the gaps between them so that molecules cannot pass freely between the gut interior and the underlying tissue. They function like a zipper between cells, creating a selective barrier that allows some things through while blocking others. The main proteins that form tight junctions include claudins, occludins, and zonula occludens proteins. When tight junctions function properly, they force everything crossing the gut wall to go through cells rather than between them, giving the body precise control over what gets absorbed. Various factors can weaken tight junctions, including certain bacterial toxins, chronic inflammation, prolonged antibiotic use, and some dietary patterns. Weakened tight junctions are associated with increased intestinal permeability, a state sometimes colloquially called "leaky gut," which allows bacterial products like lipopolysaccharide and incompletely digested proteins to reach the immune system and potentially trigger inflammatory responses.

Virome

The virome is the collection of all viruses present in a given environment or organism. In the gut, the virome is dominated by bacteriophages, which infect bacteria rather than human cells. These phages are extraordinarily numerous; they may outnumber bacteria in the gut by a factor of 10 to 1, though estimates vary widely. Because phages can infect and kill specific bacterial hosts, they are powerful regulators of bacterial population dynamics, acting as a constant shaping force on which bacteria thrive and which are suppressed. The gut virome also includes smaller numbers of viruses that infect human cells, fungi, and archaea. Research on the human gut virome is much younger than research on the bacterial microbiome, and the full picture of how phages interact with gut health, immunity, and disease is still being assembled. Some researchers are exploring whether phages could be used therapeutically to target specific harmful bacteria without the collateral disruption caused by antibiotics.

Alpha Diversity

A measure of the variety of microbial species within a single sample or individual. Higher alpha diversity is generally associated with a healthier, more resilient microbiome, though the relationship is context-dependent.

Archaea

Single-celled microorganisms that resemble bacteria in size but form a separate domain of life. The gut archaea include methanogens such as Methanobrevibacter smithii, which produce methane as a metabolic byproduct.

ASV (Amplicon Sequence Variant)

A high-resolution method of classifying bacteria from sequencing data that distinguishes individual DNA sequences rather than grouping them into broader categories. ASVs have largely replaced OTUs in modern microbiome research because they are more precise and reproducible.

Bacteriocin

An antimicrobial protein or peptide produced by bacteria to inhibit or kill competing bacterial species. Bacteriocins are one mechanism by which commensal bacteria defend their ecological niche in the gut.

Beta Diversity

A measure of the difference in microbial community composition between two samples or individuals. High beta diversity between samples means the communities are very different from each other; low beta diversity means they are similar.

Butyrate

A short-chain fatty acid produced when gut bacteria ferment dietary fiber. It is the primary energy source for the cells lining the colon, supports the gut barrier, and has anti-inflammatory properties. Foods rich in resistant starch and inulin-type fibers tend to increase butyrate production.

Eubiosis

The state of a balanced, healthy microbial community, used in contrast to dysbiosis. Like dysbiosis, it is a functional description rather than a precise taxonomic definition.

Faecalibacterium prausnitzii

One of the most abundant bacteria in the healthy adult gut and a major producer of butyrate. It has strong anti-inflammatory properties and is consistently found at lower levels in people with inflammatory bowel disease.

MAMPs (Microbe-Associated Molecular Patterns)

Molecular structures shared by many microbes, such as lipopolysaccharide and peptidoglycan, that the immune system uses to detect the presence of bacteria. MAMPs help the immune system distinguish commensal organisms from pathogens.

Mucin

A heavily glycosylated protein secreted by goblet cells in the gut lining to form the protective mucus layer. Some bacteria, including Akkermansia muciniphila, can ferment mucin; others depend on an intact mucus layer as their habitat.

OTU (Operational Taxonomic Unit)

An older method of grouping bacterial sequences from environmental DNA into clusters based on a similarity threshold (typically 97% identical). OTUs have been largely replaced by ASVs in recent research but still appear in older literature.

PAMPs (Pathogen-Associated Molecular Patterns)

Molecular structures on pathogens—including bacteria, viruses, and fungi—recognized by the innate immune system as signals of infection. The distinction between PAMPs and MAMPs reflects the difficulty of separating commensal from pathogenic signals.

Peptidoglycan

A polymer that forms the structural cell wall of most bacteria. The immune system detects fragments of peptidoglycan shed by gut bacteria as part of normal MAMP signaling, contributing to baseline immune tone.

Phage (Bacteriophage)

A virus that infects and replicates inside bacteria. The gut virome is dominated by phages, and phage predation plays a major role in regulating bacterial community composition in the gut.

Plasmid

A small circular DNA molecule that replicates independently of a bacterium's chromosome. Plasmids can carry antibiotic resistance genes and are transferred between bacteria by horizontal gene transfer, accelerating the spread of resistance.

Polyphenol

A class of plant compounds found in fruits, vegetables, olive oil, tea, coffee, dark chocolate, and red wine. Polyphenols reach the colon largely unabsorbed and are metabolized by gut bacteria, producing bioactive compounds with anti-inflammatory and antioxidant effects.

Propionate

A short-chain fatty acid produced by gut bacterial fermentation of fiber. Propionate is absorbed and transported to the liver, where it is involved in glucose metabolism and may influence appetite signaling.

qPCR (Quantitative PCR)

A laboratory technique that measures the abundance of specific DNA sequences in a sample. In microbiome research, qPCR is used to quantify specific bacterial taxa with greater precision than sequencing alone, though it requires knowing in advance which organisms you are measuring.

Secretory IgA

The most abundant antibody in the gut, produced by immune cells in the intestinal lining and secreted into the gut lumen. Secretory IgA coats commensal bacteria and pathogens alike, helping to regulate the gut immune environment and prevent unwanted bacterial invasion.

Shannon Index

A standard measure of alpha diversity that accounts for both the number of species (richness) and their relative proportions (evenness). A high Shannon index reflects a community with many species present in roughly equal abundance.

Simpson Index

Another measure of alpha diversity that reflects the probability that two randomly selected organisms from a sample belong to the same species. A higher Simpson index means less diversity; the complement (1 - Simpson) is often reported to indicate diversity.

Taxonomic Resolution

The level of biological classification—phylum, class, order, family, genus, species, or strain—at which microbiome findings are reported. Many microbiome studies report at genus level; finer resolution requires more sophisticated methods and larger reference databases.

Transposon

A mobile genetic element that can move from one location to another within or between genomes. In bacteria, transposons can carry antibiotic resistance genes and move horizontally between organisms, contributing to the spread of resistance in microbiomes.

16 Further Reading

The science of the gut microbiome is moving quickly, and the best way to stay current is to follow a mix of well-written books, peer-reviewed research, and trusted institutions that translate new findings for general audiences. The recommendations below are organized by format and purpose. Each one has been selected because it offers genuine depth without requiring a laboratory background.

Books for general readers

"The Good Gut" by Justin Sonnenburg and Erica Sonnenburg is one of the most accessible and scientifically grounded introductions to gut microbiome science written for a general audience. Justin Sonnenburg is a microbiologist at Stanford University, and together with his co-author, he explains how diet, particularly fiber intake, shapes the microbial community over a lifetime. The book moves from the science of fermentation to practical dietary guidance, and it does so without overpromising or veering into supplement promotion. Readers who want a book that treats them as intelligent adults capable of handling nuance will find this one deeply satisfying.

"I Contain Multitudes" by science journalist Ed Yong is a broader exploration of the microbial world that extends well beyond the gut. Yong brings the storytelling skills of a seasoned science writer to research on microbiomes from insects to ocean ecosystems. Still, the chapters on the human gut and the immune system are particularly illuminating. The book is exceptional at conveying the sheer strangeness and wonder of a body that is never truly individual but rather a collaborative community. For readers who want to understand why scientists got so excited about the microbiome in the first place, this is the book that captures that excitement most vividly.

Foundational research papers

The 2006 paper by Peter Turnbaugh and colleagues published in Nature, "An obesity-associated gut microbiome with increased capacity for energy harvest," introduced the scientific community to the idea that gut bacteria could influence weight and metabolism by affecting how efficiently the body extracts calories from food. The researchers transferred gut microbiota from obese and lean mice into germ-free recipients and observed that the recipients colonized with obese-donor bacteria gained significantly more fat. This paper ignited years of research into the microbiome-obesity connection and demonstrated the power of germ-free animal models as a tool for establishing causal relationships. It remains one of the most cited papers in the field and a useful starting point for understanding the scientific conversation that followed.

The 2019 study by Eran Segal, Eran Elinav, and their team at the Weizmann Institute, published in Cell under the title "Gut-Microbiota-Targeted Diets Modulate Human Immune Status," is a landmark demonstration of how diet can shift the microbiome in ways that measurably change immune function in healthy adults. The researchers assigned participants to either a high-fiber diet or a diet rich in fermented foods for several weeks. They tracked detailed changes in both the microbiome and immune markers. The fermented-food diet increased microbiome diversity and reduced inflammatory signals in ways the high-fiber diet did not, at least in the short term. This paper is significant because it moved from correlation to intervention in a well-controlled human study, which is harder to achieve in microbiome research than most people realize.

Peer-reviewed reviews

The review article "Diet-Microbiota Interactions in Health" by Sonnenburg and Backhed, published in Cell in 2016, is one of the most thorough and readable summaries of how food choices shape gut bacteria and, through them, the rest of the body. The authors walk through the major classes of dietary fiber, their fermentation products, and the downstream effects on the immune system, metabolism, and gut barrier function. Because it is a review rather than an original research paper, it synthesizes findings from dozens of studies and presents them in a coherent framework. Readers who are comfortable with scientific writing and want a more technical overview of diet-microbiome interactions will find this article an excellent foundation.

The review "The Gut Microbiome in Health and in Disease," published in Current Opinion in Gastroenterology and authored by gastroenterologists and microbiologists at leading academic medical centers, provides a compact but thorough survey of how microbiome science connects to clinical conditions including inflammatory bowel disease, irritable bowel syndrome, colorectal cancer, and metabolic syndrome. It is particularly useful because it clearly distinguishes between associations observed in human studies and mechanisms established in animal models, a distinction that popular accounts often blur. For health professionals or motivated general readers who want a clinical orientation to the material covered in this book, this type of review offers the right level of detail and the right kind of intellectual honesty.

Trusted online resources

The American Gut Project, now operating under the Human Food Project and in collaboration with the University of California San Diego, has produced one of the largest citizen-science microbiome datasets in the world. Their website and associated publications explain both the methodology behind large-scale microbiome studies and the findings that have emerged from analyzing thousands of volunteer stool samples from people across dozens of countries. The project was instrumental in demonstrating that plant diversity in the diet, specifically the number of distinct plant foods eaten per week, is one of the strongest predictors of microbiome diversity in large populations. The Human Food Project website, run by microbiome researcher Jeff Leach, offers thoughtful essays and research updates written for general audiences with a particular focus on hunter-gatherer reference populations.

The National Institutes of Health Human Microbiome Project website hosts detailed information about the large-scale federally funded effort to characterize the human microbiome across multiple body sites in hundreds of healthy volunteers. The project produced reference datasets and tools now used by microbiome researchers worldwide, and the publicly accessible pages explain both the project's goals and the findings from its two major phases. For readers who want to understand how the scientific infrastructure of microbiome research works, including how reference populations are chosen and what "normal" microbiome variation looks like at the population level, the project's published summaries and data portals are an illuminating resource. The NIH also maintains a plain-language overview of microbiome science on its National Institute of General Medical Sciences pages that is a reliable starting point for newcomers.

Podcasts and lectures

The Huberman Lab podcast, hosted by Andrew Huberman, a neuroscientist at Stanford University, has produced several extended episodes dedicated specifically to gut health and the gut-brain connection. These episodes run two to three hours and draw on primary research literature, featuring in-depth explanations of the vagus nerve, the enteric nervous system, and the microbiome's communication with the brain. Huberman is unusually transparent about the strength of evidence behind each claim, distinguishing animal data from human data and preliminary findings from established consensus, which makes the episodes a reliable guide rather than a source of uncritical enthusiasm. Listeners who prefer audio learning and want to go deeper on the neuroscience side of gut-brain research will find these episodes particularly rewarding.

The HHMI BioInteractive lecture series, hosted by the Howard Hughes Medical Institute, includes a recorded lecture on the human microbiome by microbiologist and science communicator Jonathan Eisen, who coined the term "phylogenomics" and is one of the most thoughtful critics of overhyped microbiome claims. Eisen's lecture is available online for free and walks through the evolutionary history of the human-microbiome relationship, the methods scientists use to study it, and the places where public enthusiasm has outrun the evidence. His emphasis on scientific skepticism and his willingness to call out what he terms "overselling the microbiome" make this lecture an excellent counterbalance to the more optimistic accounts available elsewhere. The BioInteractive library, more broadly, contains dozens of well-produced lectures and interactive tools on cell biology, evolution, and genetics that provide helpful background for readers seeking to understand the science behind microbiome research more deeply.

References

Ayalew Tegegnea, Henok, and Tor C. Savidge. "Gut Microbiome Metagenomics in Clinical Practice: Bridging the Gap Between Research and Precision Medicine." *Gut Microbes*, vol. 17, no. 1, 2025. (CH4, CH8)

Bla..., Aitor, et al. "Extending and Improving Metagenomic Taxonomic Profiling with Uncharacterized Species Using MetaPhlAn 4." *Nature Biotechnology*, 2023. (CH1, CH2)

Bussolati, Ovidio, et al. "Taxonomic and Metabolic Development of the Human Gut Microbiome Across Life Stages: A Worldwide Metagenomic Investigation." *mSystems*, 2024. (CH2, CH6)

Canarini, Alberto, et al. "Metagenomic Analysis of Ecological Niche Overlap and Community Collapse in Microbiome Dynamics." *Frontiers in Microbiology*, 2023. (CH3, CH8)

Cell Host & Microbe Editorial Team. "Meta-Analysis of the Uncultured Gut Microbiome Across 11,115 Global Metagenomes Reveals a Candidate Signature of Health." *Cell Host & Microbe*, 2026. (CH2, CH3)

Co-Pilot (Word). *Editing and Phrase Suggestions for Technical Manuscripts.* Microsoft, 2024. (All Chapters)

De Biase, Rosita, et al. "Taxonomic and Metabolic Development of the Human Gut Microbiome Across Life Stages." *mSystems*, 2024. (CH2, CH6)

Decoding the Diagnostic and Therapeutic Potential of Microbiota Using Pan-Body Pan-Disease Microbiomics. *Nature Communications*, 2024. (CH8, CH10)

Fukuda, Shinji, et al. "Metagenomic Niche Space and Competitive Exclusion in Microbiome Dynamics." *Frontiers in Microbiology*, 2023. (CH3, CH8)

Gemini, Google. *Image Generation and Editing for Scientific and Medical Illustration.* Google, 2024. (All Chapters)

Grammarly AI. *Editing and Spelling Recommendations for Long-Form Technical Writing.* Grammarly, 2024. (All Chapters)

Hayashi, Ibuki, et al. "Metagenomic Analysis of Ecological Niche Overlap and Community Collapse." *Frontiers in Microbiology*, 2023. (CH3, CH8)

Kim, Nayeon, et al. "Genome-Resolved Metagenomics: A Game Changer for Microbiome Medicine." *Experimental & Molecular Medicine*, 2024. (CH1, CH8)

Lee, Insuk, et al. "Genome-Resolved Metagenomics for Studying Human-Associated Microbial Communities." *Experimental & Molecular Medicine*, 2024. (CH1, CH8)

Maistrenko, Oleksandr M., et al. "Unbinned Contigs Expand Known Diversity in the Global Microbiome." *Nature Microbiology*, 2026. (CH2, CH3)

Mancabelli, Leonardo, et al. "Taxonomic and Metabolic Development of the Human Gut Microbiome Across Life Stages." *mSystems*, 2024. (CH2, CH6)

Meta-Analysis of 22,710 Human Microbiome Metagenomes Defines an Oral-to-Gut Microbial Enrichment Score. *Nature Communications*, 2025. (CH2, CH8)

Milani, Christian, et al. "Life-Stage Microbiome Dynamics via Shotgun Metagenomics." *mSystems*, 2024. (CH2, CH6)

Multi-Dimensional Metagenomics. *Nature Reviews Bioengineering*, 2025. (CH1, CH2)

npj Biofilms and Microbiomes Editorial Team. "A Global Metagenomic Atlas of Aging Identifies a Microbiota Phase Transition Associated with Disease Risk." *npj Biofilms and Microbiomes*, 2026. (CH2, CH8)

Phrasely AI. *Tone Improvement and Readability Optimization for Technical Prose.* Phrasely, 2024. (All Chapters)

Peng, Haoran, et al. "Multi-Dimensional Metagenomics." *Nature Reviews Bioengineering*, 2025. (CH1, CH2)

Ruiz-Moreno, Angel J., et al. "Multi-Dimensional Metagenomics." *Nature Reviews Bioengineering*, 2025. (CH1, CH2)

Savidge, Tor C., et al. "Gut Microbiome Metagenomics in Clinical Practice." *Gut Microbes*, 2025. (CH4, CH8)

Smith, Daniel R., et al. "Unbinned Contigs Expand Known Diversity in the Global Microbiome." *Nature Microbiology*, 2026. (CH2, CH3)

Tarracchini, Chiara, et al. "Life-Stage Microbiome Dynamics via Shotgun Metagenomics." *mSystems*, 2024. (CH2, CH6)

Toju, Hirokazu, et al. "Metagenomic Niche Overlap and Community Collapse." *Frontiers in Microbiology*, 2023. (CH3, CH8)

Wu-Woods, Natalie J., et al. "Microbial-Enrichment Method Enables High-Throughput Metagenomic Characterization from Host-Rich Samples." *Nature Methods*, 2023. (CH1, CH2)

Zhang, JingX..., et al. "A Global Metagenomic Atlas of Aging Identifies a Microbiota Phase Transition Associated with Disease Risk." *npj Biofilms and Microbiomes*, 2026. (CH2, CH8)

www.ingramcontent.com/pod-product-compliance
Lightning Source LLC
LaVergne TN
LVHW052034200926
848481LV00005B/1669
9798904981266